Artificial Neural Networks

To Sileide, Karla, Solange,
Raphael and Elen

Preface

What are artificial neural networks? What is their purpose? What are their potential practical applications? What kind of problems can they solve?

With these questions in mind, this book was written with the primary concern of answering readers with different profiles, from those interested in acquiring knowledge about architectures of artificial neural network to those motivated by its multiple applications (of practical aspect) for solving real-world problems.

This book audience is multidisciplinary, as it will be confirmed by the numerous exercises and examples addressed here. It explores different knowledge areas, such as engineering, computer science, mathematics, physics, economics, finance, statistics, and neurosciences. Additionally, it is expected that this book could be interesting for those from many other areas that have been in the focus of artificial neural networks, such as medicine, psychology, chemistry, pharmaceutical sciences, biology, ecology, geology, and so on.

Regarding the academic approach of this book and its audience, the chapters were tailored in a fashion that attempts to discuss, step-by-step, the thematic concepts, covering a broad range of technical and theoretical information. Therefore, besides meeting the professional audience's desire to begin or deepen their study on artificial neural networks and its potential applications, this book is intended to be used as a textbook for undergraduate and graduate courses, which address the subject of artificial neural networks in their syllabus.

Furthermore, the text was composed using an accessible language so it could be read by professionals, students, researchers, and autodidactics, as a straightforward and independent guide for learning basic and advanced subjects related to artificial neural networks. To this end, the prerequisites for understanding this book's content are basic, requiring only a few elementary knowledge about linear algebra, algorithms, and calculus.

The first part of this book (Chaps. 1–10), which is intended for those readers who want to begin or improve their theoretical investigation on artificial neural networks, addresses the fundamental architectures that can be implemented in several application scenarios.

The second part of this book (Chaps. 11–20) was particularly created to present solutions that comprise artificial neural networks for solving practical problems from different knowledge areas. It describes several developing details considered to achieve the described results. Such aspect contributes to mature and improve the reader's knowledge about the techniques of specifying the most appropriated artificial neural network architecture for a given application.

São Carlos, Brazil   Ivan Nunes da Silva
                                                         Danilo Hernane Spatti
                                                      Rogerio Andrade Flauzino
                                                    Luisa Helena Bartocci Liboni
                                                      Silas Franco dos Reis Alves

Organization

This book was carefully created with the mission of presenting an objective, friendly, accessible, and illustrative text, whose fundamental concern is, in fact, its didactic format. The book organization, made in two parts, along with its composition filled with more than 200 figures, eases the knowledge building for different readers' profiles. The bibliography, composed of more than 170 references, is the foundation for the themes covered in this book. The subjects are also confronted with up-to-date context.

The first part of the book (Part I), divided into ten chapters, covers the theoretical features related to the main artificial neural architectures, including Perceptron, Adaptive Linear Element (ADALINE), Multilayer Perceptron, Radial Basis Function (RBF), Hopfield, Kohonen, Learning Vector Quantization (LCQ), Counter-Propagation, and Adaptive Resonance Theory (ART). In each one of these chapters from Part I, a section with exercises was inserted, so the reader can progressively evaluate the knowledge acquired within the numerous subjects addressed in this book.

Furthermore, the chapters that address the different architectures of neural networks are also provided with sections that discuss hands-on projects, whose content assists the reader on experimental aspects concerning the problems that use artificial neural networks. Such activities also contribute to the development of practical knowledge, which may help on specifying, parameterizing, and tuning neural architectures.

The second part (Part II) addresses several applications, covering different knowledge areas, whose solutions and implementations come from the neural network architectures explored in Part I. The different applications addressed on this part of the book reflect the potential applicability of neural networks to the solution of problems from engineering and applied sciences. These applications intend to drive the reader through different modeling and mapping strategies based on artificial neural networks for solving problems of diverse nature.

Several didactic materials related to this book, including figures, exercise tips, and datasets for training neural networks (in table format), are also available at the following Website:

http://laips.sel.eesc.usp.br/

In summary, this book will hopefully create a pleasant and enjoyable reading, thus contributing to the development of a wide range of both theoretical and practical knowledge concerning the area of artificial neural networks.

Acknowledgments

The authors are immensely grateful to the many colleagues who contributed to the realization of this compilation, providing precious suggestions to help us promptly on this important and noble work.

In particular, we would like to express our thanks to the following colleagues: Alexandre C.N. de Oliveira, Anderson da Silva Soares, André L.V. da Silva, Antonio V. Ortega, Débora M.B.S. de Souza, Edison A. Goes, Ednaldo J. Ferreira, Eduardo A. Speranza, Fabiana C. Bertoni, José Alfredo C. Ulson, Juliano C. Miranda, Lucia Valéria R. de Arruda, Marcelo Suetake, Matheus G. Pires, Michelle M. Mendonça, Ricarco A.L. Rabêlo, Valmir Ziolkowski, Wagner C. Amaral, Washington L.B. Melo and Wesley F. Usida.

Contents

About the Authors

Ivan Nunes da Silva was born in São José do Rio Preto, Brazil, in 1967. He graduated in computer science and electrical engineering at the Federal University of Uberlândia, Brazil, in 1991 and 1992, respectively. He received both M.Sc. and Ph.D. degrees in Electrical Engineering from the State University of Campinas (UNICAMP), Brazil, in 1995 and 1997, respectively. Currently, he is Associate Professor at the University of São Paulo (USP). His research interests are within the fields of artificial neural networks, fuzzy inference systems, power system automation, and robotics. He is also associate editor of the International Journal on Power System Optimization and Editor-in-Chief of the Journal of Control, Automation and Electrical Systems. He has published more than 400 papers in congress proceedings, international journals, and book chapters.

Danilo Hernane Spatti was born in Araras, Brazil, in 1981. He graduated in electrical engineering from the São Paulo State University (UNESP), Brazil, in 2005. He received both M.Sc. and Ph.D. degrees in Electrical Engineering from the Uiversity of São Paulo (USP), Brazil, in 2007 and 2009, respectively. Currently, he is a Senior Researcher at the University of São Paulo. His research interests are artificial neural networks, computation complexity, systems optimization and intelligent systems.

Rogerio Andrade Flauzino was born in Franca, Brazil, in 1978. He graduated in electrical engineering and also received M.Sc. degree in electrical engineering from the São Paulo State University (UNESP), Brazil, in 2001 and 2004, respectively. He received Ph.D. degree in Electrical Engineering from the University of São Paulo (USP), Brazil, in 2007. Currently, he is Associate Professor at the University of São Paulo. His research interests include artificial neural networks, computational intelligence, fuzzy inference systems, and power systems.

Luisa Helena Bartocci Liboni was born in Sertãozinho, Brazil, in 1986. She graduated in electrical engineering from the Polytechnic School of the University of São Paulo (USP), Brazil, in 2010. She received Ph.D. degree in Electrical Engineering from the University of São Paulo (USP), Brazil, in 2016. Currently, she is Senior Researcher at the University of São Paulo. Her research interests include artificial neural networks, intelligent systems, signal processing, and nonlinear optimization.

Silas Franco dos Reis Alves was born in Marília, Brazil, in 1987. He graduated in information systems from the São Paulo State University (UNESP). He received M.Sc. degree in Mechanical Engineering from the State University of Campinas (UNICAMP) and Ph.D. degree in Electrical Engineering from the University of São Paulo (USP), Brazil, in 2011 and 2016, respectively. Currently, he is Senior Researcher at the University of São Paulo. His research interests include robotics, artificial neural networks, machine learning, intelligent systems, signal processing, and nonlinear optimization.

Part I
Architectures of Artificial Neural Networks and Their Theoretical Aspects

Chapter 1
Introduction

Building a machine or autonomous mechanism endowed with intelligence is an ancient dream of researchers from the diverse areas of sciences and engineering. Although the first articles about Artificial Neural Networks (ANN) were published more than 50 years ago, this subject began to be deeply researched on the early 90s, and still have an enormous research potential. The applications involving systems considered intelligent cover a wide range, including:

- Analysis of images acquired from artificial satellites.
- Speech and writing pattern classification.
- Face recognition with computer vision.
- Control of high-speed trains.
- Stocks forecasting on financial market.
- Anomaly identification on medical images.
- Automatic identification of credit profiles for clients of financial institutions.
- Control of electronic devices and appliances, such as washing machines, microwave ovens, freezers, coffee machines, frying machines, video cameras, and so on.

Besides these applications, the capabilities of artificial neural networks enable the solution of other kinds of problems deriving from different knowledge areas, as testified on numerous scientific journals. One example is medicine, where artificial neural networks are used for classifying and predicting cancer based on the genetic profile of a given individual (Cireşan et al. 2013; Khan et al. 2001). Other application, presented by Yan et al. (2006), proposes a decision support system, also based on artificial neural networks, for heart disease diagnosis.

The chemistry area has also registered works where artificial neural networks are used for obtaining novel polymeric compounds (Steinera et al. 2011; Zhang and Friedrich 2003). Artificial neural networks are also applied in control systems for water treatment, which involves physical and chemical nonlinear processes that are difficult to be mapped by conventional control methods (Han et al. 2011; Zhang and Stanley 1999).

© Springer International Publishing Switzerland 2017
I.N. da Silva et al., *Artificial Neural Networks*,
DOI 10.1007/978-3-319-43162-8_1

On the biology area, it is possible to find applications using artificial neural networks with the goal of identifying bat species based on their echo localization signals (biosonar) emitted during flight (Armitage and Ober 2010; Parsons and Jones 2000). Another neural approach for classifying mice species, through the sound they produced, was developed by Tian and Shang (2006).

Concerning financial and economic fields, there are also problems that are difficult to solve, mostly due to the non-linear behavior of these systems. Artificial neural networks are widely applied to such scenarios thanks to their capabilities of handling intrinsically nonlinearities (Karaa et al. 2011; Coakley and Brown 2000).

The biology field has also been beneficiated from the capabilities of artificial neural networks for extracting information, and uses them to analyze the influence of the weather on the growing of trees (Lek and Guégan 2012; Zhang et al. 2000).

The abilities of artificial neural networks for pattern classification can be observed even in the etiology field, where they are used to distinguish the diverse facial expressions conveying human emotions (Agarwal et al. 2010; Dailey and Cottrell 2002). Another application related to pattern classification is discussed by Fernandes et al. (2013), where artificial neural networks are used to classify harmonic current sources in power distribution systems.

It is possible to find in the pharmaceutical field the employment of artificial neural networks to support the formulation of novel drugs, indicating whether the medicine should be produced by microemulsion or solid dispersion methods (Deeb 2010; Mendyk and Jachowicz 2007).

On acoustics, it is also possible to find research that uses artificial neural networks for assaying the environmental acoustic impedance, a very important feature for projecting cinema rooms and environments sensitive to external noise (Hinton et al. 2012; Too et al. 2007).

The depth in which pollutants are expected to penetrate in the soil and contaminate ground water can also be estimated through artificial neural networks, providing some basis for the development of contention actions (Chowdhury et al. 2010; Tabach et al. 2007).

The food industry has also been beneficiated from the application of artificial neural networks (Argyria et al. 2010), such as those applied in classifying different varieties of tea (He et al. 2007). Another application example is presented by Silva (2007), where a neural approach was developed for processing signals from nuclear magnet resonance to classify cattle beef, allowing the identification of the sex and race of the animals. On a different application, Nazarion (2007) employed artificial neural networks combined with ultrasound techniques to classify milk regarding adulteration by the addition of fat and water.

In the automotive and aerospace industry, applications with artificial neural networks are found for assisting the mapping of processes involving estimation of control variables and project parameters. As an example of such applications, Cho et al. (2006) proposed modeling methods and control strategies for unmanned aerial vehicles. Richter et al. (2010) designed a neural architecture to perform virtual sensing of oxygen in bi-fuel vehicles. Vicente et al. (2007) proposed a neural idling speed controller for internal combustion engines. Another interesting application in the automotive area is established by Ortega and Silva (2008), where artificial neural networks are used to optimize brake light projects built with light emitting diodes.

Artificial neural networks are part of the area known as intelligent systems (connectionist systems), or computational intelligence (Jang et al. 1997; Zadeh 1992). Besides artificial neural networks, the intelligent system area includes diverse tools, such as fuzzy systems, (Pedrycz and Gomide 2007; Buckley and Siler 2004; Ross 2004), evolutionary computing (Michalewicz 1999; Dasgupta and Michalewicz 1997; Goldberg 1989), swarm intelligence (Kennedy and Eberhart 2001), artificial immunologic systems (Dasgupta 2006; Castro and Timmis 2002) and intelligent agents (D'Inverno and Luck 2004).

Additionally, today's entertainment industry, especially the cinematographic arts, has explored the subject in many science fiction movies that address the use of intelligent systems in machines and robots.

The most attractive feature of artificial neural networks, and also the source of their reputation as powerful tools for solving diverse problems, is their high capacity of mapping nonlinear systems, enabling them to learn the underlying behaviors from data acquired from such systems.

1.1 Fundamental Theory

Artificial neural networks are computational models inspired by the nervous system of living beings. They have the ability to acquire and maintain knowledge (information based) and can be defined as a set of processing units, represented by artificial neurons, interlinked by a lot of interconnections (*artificial synapses*), implemented by vectors and matrices of synaptic weights.

1.1.1 Key Features

The most relevant features concerning artificial neural applications are the following:

(a) *Adapting from experience*
    The internal parameters of the network, usually its synaptic weights, are adjusted with the examination of successive examples (patterns, samples, or measurements) related to the process behavior, thus enabling the acquisition of knowledge by experience.
(b) *Learning capability*
    Through the usage of a learning method, the network can extract the existing relationship between the several variables of the application.

(c) *Generalization capability*
Once the learning process is completed, the network can generalize the acquired knowledge, enabling the estimation of solutions so far unknown.

(d) *Data organization*
Based on innate information of a particular process, the network can organize this information, therefore enabling the clustering of patterns with common characteristics.

(e) *Fault tolerance*
Thanks to the high number of interconnections between artificial neurons, the neural network becomes a fault-tolerant system if part of its internal structure is corrupted to some degree.

(f) *Distributed storage*
The knowledge about the behavior of a particular process learned by a neural network is stored in each one of the several synapses between the artificial neurons, therefore improving the architecture robustness in case of some neurons are lost.

(g) *Facilitated prototyping*
Depending on the application particularities, most neural architectures can be easily prototyped on hardware or software, since its results, after the training process, are usually obtained with some fundamental mathematical operations.

1.1.2   Historical Overview

The first publication related to neurocomputing dates from 1943, when McCulloch and Pitts (1943) composed the first mathematical model inspired by biological neurons, resulting in the first conception of the artificial neuron.

In 1949, the first method for training artificial neural networks was proposed; it was named Hebb's rule and was based on hypothesis and observations of neuro-physiologic nature (Hebb 1949).

Many other researchers have continued the development of mathematical models based on the biological neuron, consequently generating a large number of topologies (structures) and learning algorithms. Among the different branches that emerged, the work of Frank Rosenblatt stands out. Between 1957 and 1958, Rosenblatt developed the first neurocomputer called Mark I Perceptron, crafting the basic model of the Perceptron (Rosenblatt 1958).

The Perceptron model stirred interest due to its capability of recognizing simple patterns. Widrow and Hoff (1960) developed a network called ADALINE, which is short for ADAptive LINEar Element. Later on, the MADALINE, the Multiple ADALINE, was proposed. It consisted on a network whose learning is based on the Delta rule, also known as LMS (Least Mean Square) learning method.

Following this earlier work, many researchers of that time were encouraged to conduct research in this area. However, in 1969, neurocomputing suffered a major setback with the publication of the classical book "Perceptrons: An Introduction to Computation Geometry" by Minsky and Papert (1969). The authors discussed emphatically the limitations of the neural networks of that time—which were composed of a single layer, such as the Perceptron and the ADALINE—on learning the relationship between inputs and outputs of very basic logical functions, such as XOR (exclusive or). To be more precise, that book demonstrated the impossibility of neural networks to classify patterns of nonlinearly separable classes.

Following the impact of that publication, researches on neural networks were greatly reduced, and some of the few works thereafter were: the derivation of prediction algorithms using reverse gradients (Werbos 1974), the development of the ART (Adaptive Resonance Theory) network by Grossberg (1980), the formulation of the self-organized maps (SOM) by Kohonen (1982), and the recurrent network based on energy functions proposed by Hopfield (1982). The latter is the work that brought to the artificial neural networks area its original prestige from before 1969.

Only after the end of the 1980s, supported by the work above, scientists restored their interest in this area. The definitive comeback of artificial neural networks is due to different reasons, such as the development of computers with enhanced processing and memory capabilities, the conception of more robust and efficient optimization algorithms, and finally, the novel findings about the biological nervous system. One of the fundamental works of that time was the publication of Rumelhart, Hinton and Williams' book "Parallel Distributed Processing" (Rumelhart et al. 1986), which brought to light one algorithm that allowed the adjustment of weight matrices of networks with more than a single layer. Consequently, solving the old problem of learning patterns from the XOR logical function. The proposal of this algorithm, called "backpropagation," definitely revived and motivated researches in artificial neural networks.

In recent years, together with numerous practical applications on different areas of knowledge, dozens of new and different researches have enabled theoretical advancements in artificial neural networks. Some interesting work, in particular, includes the learning algorithm based on Levenberg–Marquardt method, which fostered efficiency improvement of artificial neural networks in diverse applications (Hagan and Menhaj 1994); the artificial neural networks based on support vector machines (SVM), which can also be used for pattern classification and linear regression (Vapnik 1998); and the development of neural integrated circuits with several circuit configurations (Beiu et al. 2003).

A highly detailed description about the several other historical facts within the evolving process of artificial neural networks, since its early beginning, can be found on Haykin (2009).

1.1.3   Potential Application Areas

Artificial neural networks can be employed in several problems related to engineering and sciences. The potential application areas can be divided as follows:

(a) *Universal curve fitting (function approximation)*
    The goal is to map the functional relationship between variables (usually real numbers) of a particular system from a known set of meaningful values. These applications are as diverse as possible, and often involve mapping processes that are difficult to model using traditional methods.

(b) *Process control*
    This application category consists of identifying control actions capable of meeting quality, efficiency, and security requirements. Among the multiple available applications, neural controllers are of particular interest to robotics, airplanes, elevators, appliances, satellites, and so on.

(c) *Pattern recognition/classification*
    The purpose is to associate a given input pattern (sample) to one of the previously defined classes, as in the case of image, speech and writing recognition. In this case, the problem being addressed has a discrete and known set of possible desired outputs.

(d) *Data clustering*
    On this circumstance, the goal is to detect and identify similarities and particularities of the several input patterns to allow their grouping (clustering). Some examples, to cite a few, are applications involving automatic class identification and data mining.

(e) *Prediction system*
    The purpose of this system category is to estimate future values of a particular process, taking into account several previous samples observed in its domain. Among the known applications, it is possible to find systems for time series prediction, stock market projection, weather forecast, and so on.

(f) *System optimization*
    The goal is to minimize or maximize a cost function (objective) obeying eventual constraints to correctly map a problem. Among the optimization tasks which can benefit from artificial neural networks, the most important includes constrained optimization problems, dynamic programming, and combinational optimization.

(g) *Associative memory*
    The objective is to recover a correct pattern even when its inner elements are uncertain or inaccurate. Some examples include image processing, signal transmission, written character identification, and so forth.

1.2 Biological Neuron

The information processing performed by the human brain is carried out by biological processing components, operating in parallel, for producing proper functions, such as thinking and learning.

The fundamental cell of the central nervous system is the neuron, and its role comes down to conduct impulses (electrical stimuli originated from physical–chemical reactions) under certain operation conditions. This biological component can be divided into three main parts: dendrites, cell body (also known as "soma"), and axon.

Dendrites are composed of several thin extensions that form the dendritic tree (Fig. 1.1). The fundamental purpose of dendrites is to acquire, continuously, stimuli from several other neurons (connectors) or from the external environment, which is the case of some neurons in contact with the environment (also called sensory neurons).

The cell body is responsible for processing all the information that comes from the dendrites, to produce an activation potential that indicates if the neuron can trigger an electric impulse along its axon. It is also in the cell body where the main cytoplasmic organelles (nucleus, mitochondria, centriole, lysosome, and so forth) of the neuron can be found.

The axon is composed of a single extension whose mission is to guide the electrical impulses to other connecting neurons, or to neurons directly connected to the muscular tissue (efferent neurons). The axon termination is also composed of branches called synaptic terminals.

The synapses are the connections which enable the transfer of electric axon impulses from a particular neuron to dendrites of other neurons, as illustrated in Fig. 1.2. It is important to note that there is no physical contact between the neurons forming the synaptic junction, so the neurotransmitter elements released on the junction are in charge of weighting the transmission from one neuron to another. In

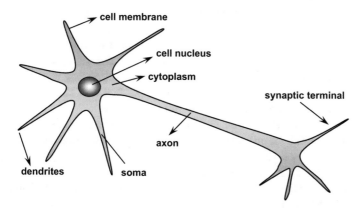

Fig. 1.1 Biological neuron

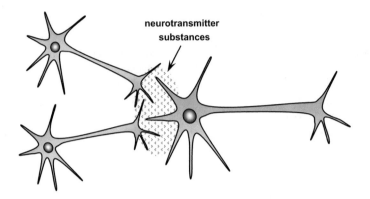

Fig. 1.2 Illustration of the synaptic connection between neurons

Property	Physical dimension
Brain mass	1.5 kg
Energy consumed by the brain	20 %
Neuron length	100 μm
Resting potential	−70 mV
Threshold potential	−55 mV
Action potential (peak)	35 mV

Table 1.1 Physical properties of the human brain and its components (in estimated values)

fact, the functionality of a neuron is dependable of its synaptic weighting, which is also dynamic and dependent on the cerebral chemistry (Hodkin and Huxley 1952).

In short, although the activities related to the biological neuron might seem very simple at first, its components, when functioning altogether, are responsible for all the processing executed and managed by the human brain. It is estimated that this biological neural network, with very eccentric features, is composed of about 100 billion (10^{11}) neurons. Each one of those is interconnected through synaptic connections (made possible by more than fifty neurotransmitter substances) to an average of 6,000 neurons, thus resulting in a total of 600 trillion synapses (Shepherd 2004). Table 1.1 presents some physical properties about the human brain (to be more precise, of an adult human) and its components.

As presented in Table 1.1, the neural membrane action potential has negative values when resting (polarized), meaning there is a larger concentration of negative ions inside the membrane than at its exterior.

When the nervous cell is stimulated (depolarized) with an impulse higher than its activation threshold (−55 mV), caused by the variation of internal concentrations of sodium (Na^+) and potassium (K^+) ions, it triggers an electrical impulse which will propagate throughout its axon with a maximum amplitude of 35 mV (Kandel et al. 2012). The stages related to variations of the action voltage within a neuron during its excitation are shown in Fig. 1.3.

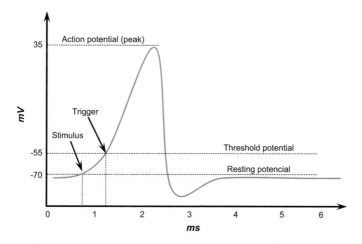

Fig. 1.3 Stages of the action potential

It is important to emphasize that the amplitude of 35 mV, the maximum value of the action voltage, is fixed and strictly satisfied for all neurons when they are stimulated, however, the signal duration in time is variable. This fact can be observed independently of the category of the neuron (connector, afferent, or efferent). As soon as the excitation process ends, the membrane will be consequently repolarized, meaning the action voltage will return to its rest voltage (−70 mV), as illustrated by Fig. 1.3.

1.3 Artificial Neuron

The artificial neural network structures were developed from known models of biological nervous systems and the human brain itself. The computational components or processing units, called artificial neurons, are simplified models of biological neurons. These models were inspired by the analysis of how a cell membrane of a neuron generates and propagates electrical impulses (Hodgkin and Huxley 1952).

The artificial neurons used in artificial neural networks are nonlinear, usually providing continuous outputs, and performing simple functions, such as gathering signals available on their inputs, assembling them according to their operational functions, and producing a response considering their innate activation functions.

The most simple neuron model that includes the main features of a biological neural network—parallelism and high connectivity—was proposed by McCulloch and Pitts (1943), and still is the most used model in different artificial neural network architectures.

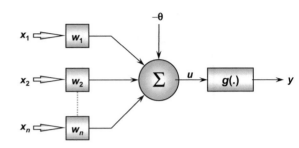

In that model, each neuron from a network can be implemented as shown in Fig. 1.4. The multiple input signals coming from the external environment (application) are represented by the set $\{x_1, x_2, x_3, \ldots, x_n\}$, analogous to the external electrical impulses gathered by the dendrites in the biological neuron.

The weighing carried out by the synaptic junctions of the network are implemented on the artificial neuron as a set of synaptic weights $\{w_1, w_2, \ldots, w_n\}$. Analogously, the relevance of each of the $\{x_i\}$ neuron inputs is calculated by multiplying them by their corresponding synaptic weight $\{w_i\}$, thus weighting all the external information arriving to the neuron. Therefore, it is possible to verify that the output of the artificial cellular body, denoted by u, is the weighted sum of its inputs.

Considering Fig. 1.4, it is possible to see that the artificial neuron is composed of seven basic elements, namely:

(a) *Input signals* $(x_1, x_2, \ldots, x_n)$ are the signals or samples coming from the external environment and representing the values assumed by the variables of a particular application. The input signals are usually normalized in order to enhance the computational efficiency of learning algorithms.

(b) *Synaptic weights* $(w_1, w_2, \ldots, w_n)$ are the values used to weight each one of the input variables, which enables the quantification of their relevance with respect to the functionality of the neuron.

(c) *Linear aggregator* (Σ) gathers all input signals weighted by the synaptic weights to produce an activation voltage.

(d) *Activation threshold or bias* (θ) is a variable used to specify the proper threshold that the result produced by the linear aggregator should have to generate a trigger value toward the neuron output.

(e) *Activation potential* (u) is the result produced by the difference between the linear aggregator and the activation threshold. If this value is positive, i.e. if $u \geq \theta$, then the neuron produces an excitatory potential; otherwise, it will be inhibitory.

(f) *Activation function* (g) whose goal is limiting the neuron output within a reasonable range of values, assumed by its own functional image.

(g) *Output signal* (y) consists on the final value produced by the neuron given a particular set of input signals, and can also be used as input for other sequentially interconnected neurons.

The two following expressions synthesize the result produced by the artificial neuron proposed by McCulloch and Pitts:

$$u = \sum_{i=1}^{n} w_i \cdot x_i - \theta \tag{1.1}$$

$$y = g(u) \tag{1.2}$$

Thus, the artificial neuron operation can be summarized by the following steps:

(i) Present a set of values to the neuron, representing the input variables.
(ii) Multiply each input of the neuron to its corresponding synaptic weight.
(iii) Obtain the activation potential produced by the weighted sum of the input signals and subtract the activation threshold.
(iv) Applying a proper activation function to limit the neuron output.
(v) Compile the output by employing the neural activation function in the activation potential.

The activation functions can be categorized into two fundamental groups, *partially differentiable functions,* and *fully differentiable functions,* when considering their complete definition domains.

1.3.1 Partially Differentiable Activation Functions

Partially differentiable activation functions are functions with points whose first order derivatives are nonexisting. The three main functions of this category are the following: step function, bipolar step function, and symmetric ramp function.

(a) *Step function (Heaviside/Hard limiter)*
   The result produced by the step function will assume unitary positive values when the neuron activation potential is greater or equal zero; otherwise, the result will be null. Thus, we have:

$$g(u) = \begin{cases} 1, & \text{if } u \geq 0 \\ 0, & \text{if } u < 0 \end{cases} \tag{1.3}$$

   The step function graphical representation is illustrated in Fig. 1.5.
(b) *Bipolar step function or Signal function (Symmetric hard limiter)*
   The result produced by this function will assume unitary positive values when the neuron activation potential is greater than zero; null value when the potential is also null; and negative unitary values when the potential is less than zero. This behavior in mathematical notation is:

Fig. 1.5 The step activation
function

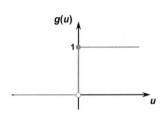

Fig. 1.6 The bipolar step
activation function

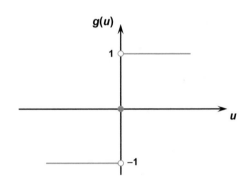

$$g(u) = \begin{cases} 1, & \text{if } u > 0 \\ 0, & \text{if } u = 0 \\ -1, & \text{if } u < 0 \end{cases} \qquad (1.4)$$

The graphical representation of this function is illustrated in Fig. 1.6.
In problems involving pattern classification, the bipolar step function can be approximated by the following expression:

$$g(u) = \begin{cases} 1, & \text{if } u \geq 0 \\ -1, & \text{if } u < 0 \end{cases} \qquad (1.5)$$

In this circumstance, another alternative is to maintain the neuron output unchanged, thus:

$$g(u) = \begin{cases} 1, & \text{if } u > 0 \\ \text{previous output}, & \text{if } u = 0 \\ -1, & \text{if } u < 0 \end{cases} \qquad (1.6)$$

(c) *Symmetric ramp function*
The values returned by this function are equal to the values of the activation potential themselves when defined within the range $[-a, a]$, and limited to the limit values otherwise. The mathematical notation for this behavior is as follows:

Fig. 1.7 The symmetric
ramp activation function

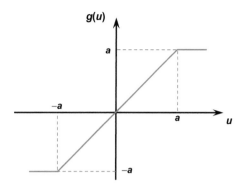

$$g(u) = \begin{cases} a, & \text{if } u > a \\ u, & \text{if } -a \le u \le a \\ -a, & \text{if } u < a \end{cases} \tag{1.7}$$

The graphical representation of this function is illustrated in Fig. 1.7.

1.3.2 Fully Differentiable Activation Functions

Fully differentiable activation functions are those whose first order derivatives exist for all points of their definition domain. The four main functions of this category, which can be employed on artificial neural networks, are the logistic function, hyperbolic tangent, Gaussian function and linear function.

(a) *Logistic function*
The output result produced by the logistic function will always assume real values between zero and one. Its mathematical expression is given by:

$$g(u) = \frac{1}{1 + e^{-\beta \cdot u}}, \tag{1.8}$$

where β is a real constant associated with the function slope in its inflection point. Figure 1.8 illustrates the behavior of this function.
Figure 1.9 shows the behavior of the logistic function when the slope parameter β changes.
From the analysis of Fig. 1.9, it is possible to conclude that the geometric format of the logistic activation function is similar to that of the step function, when β is very high, i.e., tending to infinity. However, in contrast to the step function, the logistic function is fully differentiable in its entire definition domain.

Fig. 1.8 The logistic
activation function

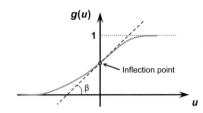

Fig. 1.9 Influence of the
parameter β in the logistic
activation function

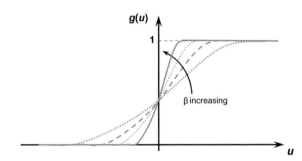

(b) *Hyperbolic tangent function*
The output result, unlike the case of the logistic function, will always assume
real values between -1 and 1, with the following mathematical expression:

$$g(u) = \frac{1 - e^{-\beta \cdot u}}{1 + e^{-\beta \cdot u}} \; , \tag{1.9}$$

where β is also associated with the slope of the hyperbolic tangent function in
its inflection point. The graphical representation of this function is illustrated
by Fig. 1.10.
Figure 1.11 also illustrates the behavior of the hyperbolic tangent function
when the parameter β changes.

Fig. 1.10 The hyperbolic
tangent activation function

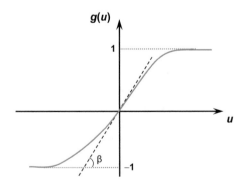

Fig. 1.11 Influence of the β parameter on the hyperbolic tangent activation function

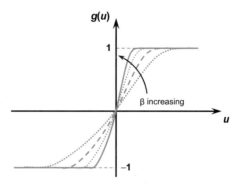

As observed in Fig. 1.11, the higher the value of β, the higher the slope the hyperbolic tangent function will have—as in the case of the logistic function —and it will approximate to the bipolar step function (signal) when the β value is very high.

It is important to note that both logistic and hyperbolic tangent functions belong to a family of functions called sigmoidal.

(c) *Gaussian function*

In the case of Gaussian activation functions, the neuron output will produce equal results for those activation potential values $\{u\}$ placed at the same distance from its center (average). The curve is symmetric to this center and the Gaussian function is given by:

$$g(u) = e^{-\frac{(u-c)^2}{2\sigma^2}}, \qquad (1.10)$$

where c is the parameter that defines the center of the Gaussian function and σ denotes the associated standard deviation, that is, how scattered (dispersed) is the curve in relation to its center. The graphical representation of this function is illustrated by Fig. 1.12.

Fig. 1.12 The Gaussian activation function

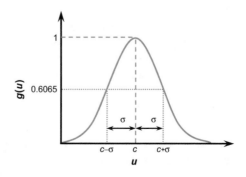

Fig. 1.13 The linear
activation function

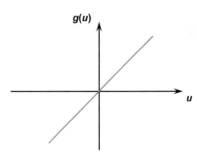

It is possible to observe on this figure that the standard deviation parameter $\{\sigma\}$ is directly associated with the inflection points of the Gaussian function, with σ^2 indicating its variance.

(d) *Linear function*

The linear activation function, or identity function, produces output results equal to the activation potential $\{u\}$, having its mathematical expression given by:

$$g(u) = u \qquad (1.11)$$

The graphical representation of this function is illustrated in Fig. 1.13.

One application of the linear activation functions is in artificial neural networks performing universal curve fitting (function approximation), to map the behavior of the input/output variables of a particular process, as it is discussed in Sect. 5.4.

1.4   Performance Parameters

To relate the operation of both artificial and biological neurons, Table 1.2 presents a comparison between their features of performance.

It is possible to observe that the processing time of artificial neurons is lower than that of biological neurons. On the other hand, the cerebral processing is countlessly faster, in most cases, than any artificial neural network, since neurons from biological neural networks operate with high degree of parallelism. Neurons from artificial neural networks have very limited parallelism capabilities, because most computers are built with sequential machines (Haykin 2009; Faggin 1991).

Table 1.2 Comparative chart between artificial and biological neurons

Parameter	Artificial neuron	Biological neuron
Energetic efficiency (operation/second)	10^{-6} J	10^{-16} J
Processing time (operation/neuron)	10^{-9} s (clock on the order GHz)	10^{-3} s
Processing mechanism	Usually sequential	Usually parallel

The speed parameter of artificial neural networks is essentially related to the number of operations per second performed by computers. Considering a clock on the order of gigahertz, the processing period of artificial neurons are in the magnitude of nanoseconds.

1.5 Exercises

1. Explain how an artificial neuron operates.
2. Describe what are the main goals of activation functions.
3. Make an analogy between the elements composing artificial and biological neurons.
4. Write about the importance of the activation threshold (or bias).
5. Thinking about the features of artificial neural networks, explain what is learning from experience and generalization capability.
6. Write about the main mathematical features which can be verified on both the logistic and hyperbolic tangent activation functions.
7. Find the analytical expressions of the first order derivatives of the logistic and hyperbolic tangent.
8. For a particular problem, it is possible to use the logistic or hyperbolic functions as the activation function. Regarding hardware implementation, write about the eventual features to be considered for selecting one of them.
9. Given that individual operation on artificial neurons are executed faster when compared to biological neurons, explain why many tasks performed by the human brain produce results faster than a microcomputer.
10. What are the main categories of problems which can be addressed by artificial neural networks?

Chapter 2
Artificial Neural Network Architectures and Training Processes

2.1 Introduction

The architecture of an artificial neural network defines how its several neurons are arranged, or placed, in relation to each other. These arrangements are structured essentially by directing the synaptic connections of the neurons.

The topology of a given neural network, within a particular architecture, can be defined as the different structural compositions it can assume. In other words, it is possible to have two topologies belonging to the same architecture, where the first topology is composed of 10 neurons, and the second is composed of 20 neurons. Moreover, one can consist of neurons with logistic activation function, while the other one can consist of neurons with the hyperbolic tangent as the activation function.

On the other hand, training a particular architecture involves applying a set of ordinated steps to adjust the weights and thresholds of its neurons. Hence, such adjustment process, also known as learning algorithm, aims to tune the network so that its outputs are close to the desired values.

2.2 Main Architectures of Artificial Neural Networks

In general, an artificial neural network can be divided into three parts, named layers, which are known as:

(a) *Input layer*

This layer is responsible for receiving information (data), signals, features, or measurements from the external environment. These inputs (samples or patterns) are usually normalized within the limit values produced by activation functions. This normalization results in better numerical precision for the mathematical operations performed by the network.

© Springer International Publishing Switzerland 2017
I.N. da Silva et al., *Artificial Neural Networks*,
DOI 10.1007/978-3-319-43162-8_2

(b) *Hidden, intermediate, or invisible layers*
These layers are composed of neurons which are responsible for extracting patterns associated with the process or system being analyzed. These layers perform most of the internal processing from a network.

(c) *Output layer*
This layer is also composed of neurons, and thus is responsible for producing and presenting the final network outputs, which result from the processing performed by the neurons in the previous layers.

The main architectures of artificial neural networks, considering the neuron disposition, as well as how they are interconnected and how its layers are composed, can be divided as follows: (i) single-layer feedforward network, (ii) multilayer feedforward networks, (iii) recurrent networks and (iv) mesh networks.

2.2.1 Single-Layer Feedforward Architecture

This artificial neural network has just one input layer and a single neural layer, which is also the output layer. Figure 2.1 illustrates a simple-layer feedforward network composed of *n* inputs and *m* outputs.

The information always flows in a single direction (thus, unidirectional), which is from the input layer to the output layer. From Fig. 2.1, it is possible to see that in networks belonging to this architecture, the number of network outputs will always coincide with its amount of neurons. These networks are usually employed in pattern classification and linear filtering problems.

Among the main network types belonging to feedforward architecture are the Perceptron and the ADALINE, whose learning algorithms used in their training processes are based respectively on Hebb's rule and Delta rule, as it will be discussed in the next chapters.

Fig. 2.1 Example of a single-layer feedforward network

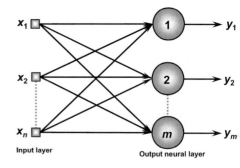

2.2.2   Multiple-Layer Feedforward Architectures

Differently from networks belonging to the previous architecture, feedforward networks with multiple layers are composed of one or more hidden neural layers (Fig. 2.2). They are employed in the solution of diverse problems, like those related to function approximation, pattern classification, system identification, process control, optimization, robotics, and so on.

Figure 2.2 shows a feedforward network with multiple layers composed of one input layer with n sample signals, two hidden neural layers consisting of n_1 and n_2 neurons respectively, and, finally, one output neural layer composed of m neurons representing the respective output values of the problem being analyzed.

Among the main networks using multiple-layer feedforward architectures are the Multilayer Perceptron (MLP) and the Radial Basis Function (RBF), whose learning algorithms used in their training processes are respectively based on the generalized delta rule and the competitive/delta rule. These concepts will be addressed in the next chapters.

From Fig. 2.2, it is possible to understand that the amount of neurons composing the first hidden layer is usually different from the number of signals composing the input layer of the network. In fact, the number of hidden layers and their respective amount of neurons depend on the nature and complexity of the problem being mapped by the network, as well as the quantity and quality of the available data about the problem. Nonetheless, likewise for simple-layer feedforward networks, the amount of output signals will always coincide with the number of neurons from that respective layer.

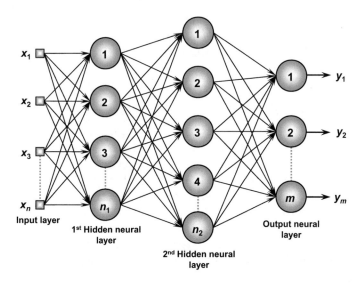

Fig. 2.2   Example of a feedforward network with multiple layers

Fig. 2.3 Example of a
recurrent network

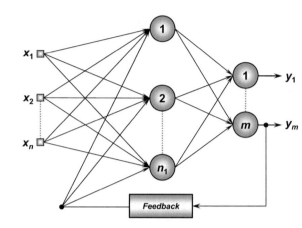

2.2.3 Recurrent or Feedback Architecture

In these networks, the outputs of the neurons are used as feedback inputs for other
neurons. The feedback feature qualifies these networks for dynamic information
processing, meaning that they can be employed on time-variant systems, such as
time series prediction, system identification and optimization, process control, and
so forth.

Among the main feedback networks are the Hopfield and the Perceptron with
feedback between neurons from distinct layers, whose learning algorithms used in
their training processes are respectively based on energy function minimization and
generalized delta rule, as will be investigated in the next chapters.

Figure 2.3 illustrates an example of a Perceptron network with feedback, where
one of its output signals is fed back to the middle layer.

Thus, using the feedback process, the networks with this architecture produce
current outputs also taking into consideration the previous output values.

2.2.4 Mesh Architectures

The main features of networks with mesh structures reside in considering the spatial
arrangement of neurons for pattern extraction purposes, that is, the spatial localization of the neurons is directly related to the process of adjusting their synaptic
weights and thresholds. These networks serve a wide range of applications and are
used in problems involving data clustering, pattern recognition, system optimization, graphs, and so forth.

The Kohonen network is the main representative of mesh architectures, and its
training is performed through a competitive process, as will be described in the
following chapters. Figure 2.4 illustrates an example of the Kohonen network
where its neurons are arranged within a two-dimensional space.

Fig. 2.4 Structure of a mesh network

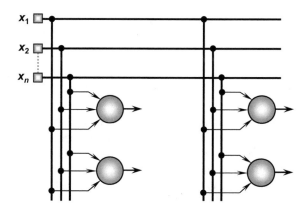

From Fig. 2.4, it is possible to verify that in this network category, the several input signals are read by all neurons within the network.

2.3   Training Processes and Properties of Learning

One of the most relevant features of artificial neural networks is their capability of learning from the presentation of samples (patterns), which expresses the system behavior. Hence, after the network has learned the relationship between inputs and outputs, it can generalize solutions, meaning that the network can produce an output which is close to the expected (or desired) output of any given input values.

Therefore, the training process of a neural network consists of applying the required ordinated steps for tuning the synaptic weights and thresholds of its neurons, in order to generalize the solutions produced by its outputs.

The set of ordinated steps used for training the network is called learning algorithm. During its execution, the network will thus be able to extract discriminant features about the system being mapped from samples acquired from the system.

Usually, the complete set containing all available samples of the system behavior is divided into two subsets, which are called training subset and test subset. The training subset, composed of 60–90 % of random samples from the complete set, will be used essentially in the learning process. On the other hand, the test subset, which is composed of 10–40 % from the complete sample set, will be used to verify if the network capabilities of generalizing solutions are within acceptable levels, thus allowing the validation of a given topology. Nonetheless, when dimensioning these subsets, statistical features of the data must also be considered.

During the training process of artificial neural networks, each complete presentation of all the samples belonging to the training set, in order to adjust the synaptic weights and thresholds, will be called training epoch.

2.3.1 Supervised Learning

The supervised learning strategy consists of having available the desired outputs for a given set of input signals; in other words, each training sample is composed of the input signals and their corresponding outputs. Henceforth, it requires a table with input/output data, also called attribute/value table, which represents the process and its behavior. It is from this information that the neural structures will formulate "hypothesis" about the system being learned.

In this case, the application of supervised learning only depends on the availability of that attribute/value table, and it behaves as if a "coach" is teaching the network what is the correct response for each sample presented for its input.

The synaptic weights and thresholds of the network are continually adjusted through the application of comparative actions, executed by the learning algorithm itself, that supervise the discrepancy between the produced outputs with respect to the desired outputs, using this difference on the adjustment procedure. The network is considered "trained" when this discrepancy is within an acceptable value range, taking into account the purposes of generalizing solutions.

In fact, the supervised learning is a typical case of pure inductive inference, where the free variables of the network are adjusted by knowing a priori the desired outputs for the investigated system.

Donald Hebb proposed the first supervised learning strategy in 1949, inspired by neurophysiological observations (Hebb 1949).

2.3.2 Unsupervised Learning

Different from supervised learning, the application of an algorithm based on unsupervised learning does not require any knowledge of the respective desired outputs.

Thus, the network needs to organize itself when there are existing particularities between the elements that compose the entire sample set, identifying subsets (or clusters) presenting similarities. The learning algorithm adjusts the synaptic weights and thresholds of the network in order to reflect these clusters within the network itself.

Alternatively, the network designer can specify (a priori) the maximum quantity of these possible clusters, using his/her knowledge about the problem.

2.3.3 Reinforcement Learning

Methods based on reinforcement learning are considered a variation of supervised learning techniques, since they continuously analyze the difference between the response produced by the network and the corresponding desired output (Sutton and Barto 1998). The learning algorithms used on reinforcement learning adjusts the internal neural parameters relying on any qualitative or quantitative information acquired through the interaction with the system (environment) being mapped, using this information to evaluate the learning performance.

The network learning process is usually done by trial and error because the only available response for a given input is whether it was satisfactory or unsatisfactory. If satisfactory, the synaptic weights and thresholds are gradually incremented to reinforce (reward) this behavioral condition involved with the system.

Several learning algorithms used by reinforcement learning are based on stochastic methods that probabilistically select the adjustment actions, considering a finite set of possible solutions that can be rewarded if they have chances of generating satisfactory results. During the training process, the probabilities associated with action adjustment are modified to enhance the network performance (Tsoukalas and Uhrig 1997).

This adjustment strategy has some similarities to some dynamic programming techniques (Bertsekas and Tsitsiklis 1996; Watkins 1989).

2.3.4 Offline Learning

In offline learning, also named batch learning, the adjustments on the weight vectors and thresholds of the network are performed after all the training set is presented, since each adjustment step takes into account the number of errors observed within the training samples with respect to the desired values for their outputs.

Therefore, networks using offline learning requires, at least, one training epoch for executing one adjustment step on their weights and thresholds. Hence, all training samples must be available during the whole learning process.

2.3.5 Online Learning

Opposite to offline learning, in online learning, the adjustments on the weights and thresholds of the network are performed after presenting each training sample. Thus, after executing the adjustment step, the respective sample can be discarded.

Online learning with this configuration is usually used when the behavior of the system being mapped changes rapidly, thus the adoption of offline learning is almost impractical because the samples used at a given moment may no more represent the system behavior in posterior moments.

However, since patterns are presented one at a time, weight and threshold adjustment actions are well located and punctual, and they reflect a given behavioral circumstance of the system. Therefore, the network will begin to provide accurate responses after presenting a significant number of samples (Reed and Marks II, 1999).

2.4   Exercises

1. Write about the advantages and disadvantages involved with online and offline learning.
2. Consider an application with four inputs and two outputs. The designers of this application state that the feedforward network to be developed must present exactly four neurons in the first hidden layer. Discuss about the pertinence of this information.
3. Relating to the previous exercise, cite some factors that influence the determination of the hidden layers number of a multiple layer feedforward network.
4. What are the eventual structural differences observed between recurrent networks and feedforward networks.
5. In what application categories the employment of recurrent neural networks is essential?
6. Draw a block diagram illustrating how the supervised training works.
7. Write about the concepts of training methods and learning algorithms, further explaining the concept of training epoch.
8. What are the main differences between supervised and unsupervised training methods?
9. What are the main differences between supervised and reinforcement learning methods?
10. Considering a specific application, explain what performance criterion could be used for adjusting the weights and thresholds of a network using reinforcement learning method.

Chapter 3
The Perceptron Network

3.1 Introduction

The Perceptron, created by Rosenblatt, is the simplest configuration of an artificial neural network ever created, whose purpose was to implement a computational model based on the retina, aiming an element for electronic perception. One application of the Perceptron was to identify geometric patterns.

Figure 3.1 illustrates one example of the initial concept of the Perceptron element, in which signals from photocells, used to map geometric patterns, were pondered by tunable resistors, which could be adjusted during the training process. After that, an additive element would combine all pondered signals. In this way, the Perceptron could recognize different geometric patterns, such as numbers and letters.

The simplicity of the Perceptron network is due to its condition of being constituted by just one neural layer, hence having a single artificial neuron in this layer.

Figure 3.2 illustrates one Perceptron network composed of n input signals, representing the problem being analyzed, and just one output, as it is composed of a single neuron.

Even though the Perceptron was a simple network, at the time it was proposed, it attracted several researchers who aspired to investigate this promising field, receiving special attention of the scientific community.

Considering the computational aspects, which are addressed in Sect. 3.2, it is possible to observe in Fig. 3.2 that the activation threshold (or bias) value θ assumed (without any loss of interpretability) the value of the pondering variable $\{w_0\}$, thus having the negative unitary value as its respective input.

Based on the discussion of the previous chapter, the Perceptron network belongs to the class of single-layer feedforward architectures, because the information that flows in its structure is always from the input layer to the output layer, without any feedback from the output produced by its single output neuron.

© Springer International Publishing Switzerland 2017
I.N. da Silva et al., *Artificial Neural Networks*,
DOI 10.1007/978-3-319-43162-8_3

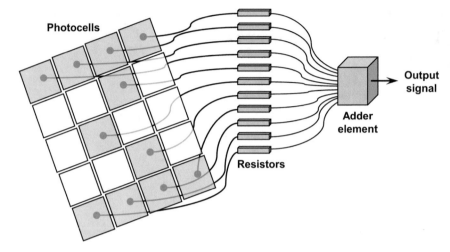

Fig. 3.1 Illustrative model of the Perceptron for pattern recognition

3.2   Operating Principle of the Perceptron

Just like the simplicity of the Perceptron structure, its operating principle is also simple.

From the analysis of Fig. 3.2, it is possible to see that each one of the x_i inputs is initially pondered by synaptic weights in order to quantify the importance of the inputs on the functional goals of the neuron, whose purpose is to map the input/output behavior of the process in question.

In sequence, the value resulting from the composition of all inputs pondered by weights, added to the activation threshold θ, is used as an argument for the activation function, whose value is the output y produced by the Perceptron.

In mathematical notation, the inner processing performed by the Perceptron can be described by the following expressions:

Fig. 3.2 Illustration of the Perceptron network

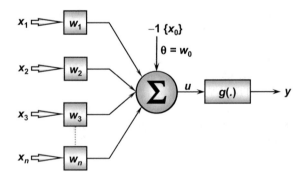

$$\begin{cases} u = \sum\limits_{i=1}^{n} w_i \cdot x_i - \theta & (3.1) \\ y = g(u), & (3.2) \end{cases}$$

where x_i is a network input, w_i is the weight (pondering) associated with the ith input, θ is the activation threshold (or bias), $g(.)$ is the activation function and u is the activation potential.

Due to their structural features, the activation functions usually used on the Perceptron are the step and the bipolar step functions, presented in Sect. 1.3.1. Thus, independently of the activation function in use, only two values can be produced by the network output, that is, 0 or 1 for the step function, and -1 or 1 if the bipolar step function is used.

The x_i inputs can assume any numeric value and depend on the problem being mapped by the Perceptron. In practice, techniques for input normalization, considering the numerical range of the adopted activation function, are also used to improve the computational performance of the training process.

To summarize, Table 3.1 presents the features of parameters related to the operation dynamics of the Perceptron.

The adjustment of weights and threshold in the Perceptron is made through supervised training, meaning that the respective desired output (response) must exist for each sample of the input signal. Since the Perceptron is usually used on pattern recognition problems, and considering that its output can assume just two possible values, then each output is associated with one of the two classes that are being identified.

More specifically, consider a problem involving the classification of input signals into two possible classes, namely *Class A* and *Class B*. It would be possible (assuming the use of the bipolar step activation function) to attribute value -1 to represent the samples belonging to *Class A*, while value 1 would be used for those of *Class B*, or vice versa.

Table 3.1 Characteristic features of the Perceptron parameters

Parameter	Variable	Type
Inputs	x_i (ith input)	Real or binary (from the external environment)
Synaptic weights	w_i (associated with x_i)	Real (initialized with random values)
Threshold (bias)	θ	Real (initialized with random values)
Output	y	Binary
Activation function	$g(.)$	Step or bipolar step function
Training process	–	Supervised
Learning rule	–	Hebb's rule

3.3 Mathematical Analysis of the Perceptron

From the mathematical analysis of the Perceptron, and by considering the signal activation function, it becomes possible to verify that such network can be considered a typical case of a linear discriminator. To demonstrate this scenario, consider a Perceptron with only two inputs, as illustrated in Fig. 3.3.

In mathematical notation, the Perceptron output, which uses the bipolar step activation function, defined by (1.5), is given by:

$$y = \begin{cases} 1, & \text{if } \sum w_i \cdot x_i - \theta \geq 0 \Leftrightarrow w_1 \cdot x_1 + w_2 \cdot x_2 - \theta \geq 0 \\ -1, & \text{if } \sum w_i \cdot x_i - \theta < 0 \Leftrightarrow w_1 \cdot x_1 + w_2 \cdot x_2 - \theta < 0 \end{cases} \tag{3.3}$$

As the inequalities in (3.3) are represented by linear equations, the classification boundary for this case (Perceptron with two inputs) will be a straight line given by:

$$w_1 \cdot x_1 + w_2 \cdot x_2 - \theta = 0 \tag{3.4}$$

Thus, it is possible to conclude that the Perceptron behaves as a pattern classifier whose purpose is to divide linearly separable classes. For the Perceptron with two inputs, Fig. 3.4 illustrates a line positioned on the decision (separating) boundary.

Fig. 3.3 Perceptron constituted by two inputs

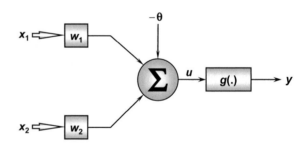

Fig. 3.4 Illustration of the decision boundary (of a neuron with two inputs)

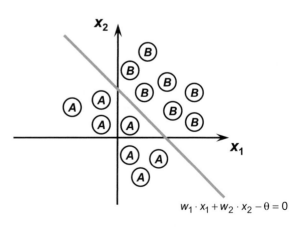

Fig. 3.5 Illustration of a nonlinear decision boundary

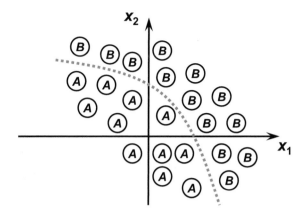

In short, in the scenario presented in Fig. 3.4, the Perceptron can separate two linearly separable classes: when its output is −1, it means that the patterns (*Class A*) are located bellow the decision boundary (straight line); otherwise, when the output is 1, it means that the patterns (*Class B*) are above this border. Figure 3.5 illustrates a configuration where the classes are not linearly separable, meaning that a single line is unable to separate the two classes being analyzed.

Considering a scenario where the Perceptron is composed of three inputs (three dimensions), the decision boundary will be represented by a plane; in higher dimensions, such boundaries will be hyperplanes.

Finally, one can conclude that the condition for a single-layer Perceptron to be applied as a pattern classifier is that the classes of the problem being mapped must be linearly separable. This conditional principle was called Perceptron convergence theorem (Minsky and Papert 1969).

3.4 Training Process of the Perceptron

The adjustment of Perceptron's weights and thresholds, in order to classify patterns that belong to one of the two possible classes, is performed by the use of Hebb's learning rule (Hebb 1949).

In short, if the output produced by the Perceptron coincides with the desired output, its synaptic weights and threshold remain unchanged (inhibitory condition); otherwise, in the case the produced output is different from the desired value, then its synaptic weights and threshold are adjusted proportionally to its input signals (excitatory condition). This process is repeated sequentially for all training samples until the output produced by the Perceptron is similar to the desired output of all samples. In mathematical notation, the rules for adjusting the synaptic weight w_i and threshold θ can be expressed, respectively, by the following equations:

$$w_i^{\text{current}} = w_i^{\text{previous}} + \left(d^{(k)} - y\right) \cdot x_i^{(k)} \tag{3.5}$$

$$\theta^{\text{current}} = \theta^{\text{previous}} + \left(d^{(k)} - y\right) \cdot (-1) \tag{3.6}$$

However, for computational implementation, it is more convenient to approach the previous equations in their vector form. As the same adjustment rule is applied to both the synaptic weights and threshold, it is possible to insert the threshold value θ within the synaptic weight vector. In fact, the threshold value is also a variable that can be adjusted to perform the training of the Perceptron. Thus, Eqs. (3.3) and (3.4) can be represented by a single vector expression given by:

$$\boldsymbol{w}^{\text{current}} = \boldsymbol{w}^{\text{previous}} + \eta \cdot \left(d^{(k)} - y\right) \cdot \boldsymbol{x}^{(k)} \tag{3.7}$$

In algorithmic notation, the inner processing performed by the Perceptron can be described by the following expression:

$$\boldsymbol{w} \leftarrow \boldsymbol{w} + \eta \cdot \left(d^{(k)} - y\right) \cdot \boldsymbol{x}^{(k)}, \tag{3.8}$$

where:

$\boldsymbol{w} = \begin{bmatrix} \theta & w_1 & w_2 & \cdots & w_n \end{bmatrix}^T$ is the vector containing the threshold and weights;
$\boldsymbol{x}^{(k)} = \begin{bmatrix} -1 & x_1^{(k)} & x_2^{(k)} & \cdots & x_n^{(k)} \end{bmatrix}^T$ is the kth training sample;
$d^{(k)}$ is the desired value for the kth training sample;
y is the output produced by the Perceptron; and
η is a constant that defines the learning rate of the Perceptron.
The learning rate η determines how fast the training process will take to its convergence (stabilization). The choice of η should be done carefully to avoid instabilities in the training process, and it is usually defined within the range $0 < \eta < 1$.

In order to clarify the notation used for vector $\boldsymbol{x}^{(k)}$ representing the kth training sample, as well as its respective desired value $d^{(k)}$, suppose a problem to be mapped by a Perceptron with three inputs $\{x_1, x_2, x_3\}$. The training set is provided by only four samples, composed of the following values $\Omega^{(x)} = \{[0.1\ 0.4\ 0.7]; [0.3\ 0.7\ 0.2]; [0.6\ 0.9\ 0.8]; [0.5\ 0.7\ 0.1]\}$. Considering that the respective output values for each of these samples are given by $\Omega^{(d)} = \{[1]; [-1]; [-1]; [1]\}$, then the following matrix notation can be adopted to represent the problem:

$$\Omega^{(\boldsymbol{x})} = \begin{matrix} & \begin{matrix} \boldsymbol{x}^{(1)} & \boldsymbol{x}^{(2)} & \boldsymbol{x}^{(3)} & \boldsymbol{x}^{(4)} \end{matrix} \\ \begin{matrix} x_0 \\ x_1 \\ x_2 \\ x_3 \end{matrix} & \begin{bmatrix} -1 & -1 & -1 & -1 \\ 0.1 & 0.3 & 0.6 & 0.5 \\ 0.4 & 0.7 & 0.9 & 0.7 \\ 0.7 & 0.2 & 0.8 & 0.1 \end{bmatrix} \end{matrix} ; \quad \Omega^{(\boldsymbol{d})} = \begin{matrix} \begin{matrix} d^{(1)} & d^{(2)} & d^{(3)} & d^{(4)} \end{matrix} \\ \begin{bmatrix} 1 & -1 & -1 & 1 \end{bmatrix} \end{matrix}$$

Alternatively, it is possible to extract from those matrices each vector $x^{(k)}$ with its respective $d^{(k)}$ value, to present to the network each training sample with the following notation:

$$
\begin{aligned}
x^{(1)} &= \begin{bmatrix} -1 & 0.1 & 0.4 & 0.7 \end{bmatrix}^T; \quad \text{with} \quad d^{(1)} = 1 \\
x^{(2)} &= \begin{bmatrix} -1 & 0.3 & 0.7 & 0.2 \end{bmatrix}^T; \quad \text{with} \quad d^{(2)} = -1 \\
x^{(3)} &= \begin{bmatrix} -1 & 0.6 & 0.9 & 0.8 \end{bmatrix}^T; \quad \text{with} \quad d^{(3)} = -1 \\
x^{(4)} &= \begin{bmatrix} -1 & 0.5 & 0.7 & 0.1 \end{bmatrix}^T; \quad \text{with} \quad d^{(4)} = 1
\end{aligned}
$$

However, when considering the computational implementation of this problem, the usage of indexed variables for storing data is much more appropriate for handling the sample indexes.

Thus, the step-by-step sequence for the Perceptron training is given by the following pseudocode.

Begin {PERCEPTRON Algorithm – Training Phase}

<1> Obtain the set of training samples {$x^{(k)}$};

<2> Associate each desired output {$d^{(k)}$} to each sample;

<3> Initialize vector w with small random values;

<4> Specify the learning rate {η};

<5> Initialize the epoch counter {*epoch* ← 0};

<6> Repeat the following instructions:

   <6.1> *error* ← "none";

   <6.2> For all training samples {$x^{(k)}$, $d^{(k)}$}, do:

   <6.2.1> $u \leftarrow w^T \cdot x^{(k)}$;

   <6.2.2> $y \leftarrow \text{signal}(u)$;

   <6.2.3> If $y \neq d^{(k)}$

   <6.2.3.1> then $\begin{cases} w \leftarrow w + \eta \cdot (d^{(k)} - y) \cdot x^{(k)} \\ error \leftarrow \text{"existent"} \end{cases}$

   <6.3> *epoch* ← *epoch* + 1;

   Until: *error* ← "none"

End {PERCEPTRON Algorithm – Training Phase}

By analyzing the algorithm involved with the training process of the Perceptron, it is possible to verify that the variable *epoch* is responsible for counting the number of training epochs, that is, how many times is necessary to present all inputs from the training set in order to adjust the weight vector $\{w\}$. The network is considered trained (adjusted) when no error is found between the desired values and those produced by the network outputs.

Once the network is trained, it is then ready to proceed with the pattern classification task when new samples are presented to its inputs. Therefore, the instructions required to get the Perceptron to work, after concluding its training, are synthesized on the following algorithm.

Begin {PERCEPTRON Algorithm – Operation Phase}

<1> Obtain one sample to be classified $\{x\}$;

<2> Use the vector w adjusted during training;

<3> Execute the following instructions:

    <3.1> $u \leftarrow w^T \cdot x$;

    <3.2> $y \leftarrow \text{signal}(u)$;

    <3.3> If $y = -1$

        <3.3.1> then: sample $x \in$ *{Class A}*

    <3.4> If $y = 1$

        <3.4.1> then: sample $x \in$ *{Class B}*

End {PERCEPTRON Algorithm – Operation Phase}

Therefore, it is possible to understand that the Perceptron training process tends to move the classification hyperplane continuously until it meets a decision boundary that allows the separation of both classes.

Figure 3.6 illustrates the Perceptron training process aiming to reach the decision boundary. For visualization purposes, this network has only two inputs $\{x_1$ and $x_2\}$.

After the first training epoch (1), the hyperplane is still quite far from the separating boundary of classes A and B, but this distance tends to decrease as the number of epochs increases.

In consequence, when the Perceptron converges, it means that the boundary was reached, and, thus, all the outputs produced by the Perceptron from this point on are equal to the desired.

The analysis of Fig. 3.6 shows the possibility of having different lines that successfully separate both classes involved with the problem. Figure 3.7 illustrates one set with eventual lines that are also capable of separating such classes.

Also, this pattern classification solution using Perceptron also demonstrates that the separability straight line produced by its training is not unique, and thus the number of epochs may vary in these cases.

Fig. 3.6 Illustration of the Perceptron training process

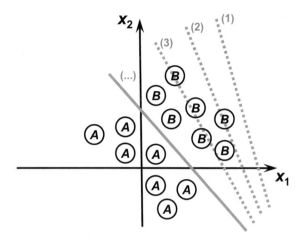

Fig. 3.7 Illustration of a set of lines capable of separating classes A and B

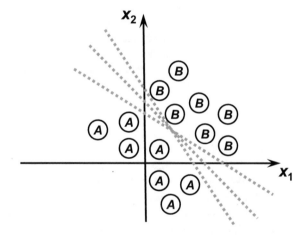

Next, some practical features involving the Perceptron training process are presented.

(a) The network will diverge if the problem is nonlinearly separable. The strategy for this scenario is to limit the training process by a maximum number of epochs.

(b) When the decision boundary between two classes is too narrow, its training process can imply instability. In such cases, by assuming a small learning rate value $\{\eta\}$, this instability might be mitigated.

(c) The amount of epochs required for the training process convergence varies according to the initial values attributed to the weight vector $\{w\}$, as well as to the initial arrangement of training samples and the specified value for the learning rate $\{\eta\}$.

(d) The closer the decision surface is to the separating boundary, the fewer epochs are usually required for the Perceptron convergence.

(e) The normalization of inputs into appropriated domains provides performance enhancements on the training process.

3.5   Exercises

1. Explain how the Hebb's rule is executed within the context of the Perceptron learning algorithm.

2. Illustrate with graphs how the instability problem can occur during the Perceptron convergence when using inappropriate values for the learning rate.

3. Explain why the Perceptron can only classify patterns whose separating boundary between classes is linear.

4. Concerning computational implementation issues, describe the importance of considering the activation threshold $\{\theta\}$ as one of the elements of the weight vector $\{w\}$.

5. Consider a problem of pattern classification with no information whether its classes are linearly separable. Elaborate one strategy to verify if the Perceptron can be applied to such problem.

6. Two designers of different institutes are applying a Perceptron network for mapping the same pattern classification problem. Is it possible to affirm that both networks will converge with the same number of epochs?

7. Consider that the networks from the previous exercise are already trained. Explain if the results produced by both networks are the same when a set with 10 new samples is presented to them.

8. A problem of pattern classification is linearly separable and has 50 samples. In a given training epoch, it was observed that the network was not producing the desired output for only one sample. Is it necessary to present all the 50 samples on the next training epoch?

9. Consider a problem of pattern classification composed of two inputs $\{x_1$ and $x_2\}$ whose training set is composed of the following:

x_1	x_2	Class
0.75	0.75	A
0.75	0.25	B
0.25	0.75	B
0.25	0.25	A

Is it possible to apply the Perceptron for solving this problem?

10. Explain in details what possible limitations the Perceptron would meet if its activation threshold was null $\{\theta = 0\}$.

3.6 Practical Work

The analysis of a fractional distillation process for petrol revealed that a given oil could be classified into two purity classes $\{P_1 \text{ and } P_2\}$ from the measurement of three variables $\{x_1, x_2 \text{ and } x_3\}$, which represent some physicochemical properties of the oil. The team of engineers and scientists proposed the application of a Perceptron network to perform the automatic classification of both classes.

Thus, based on the information gathered about the process, the team composed the training set presented in Appendix A, using a convention where the value -1 indicates oil belonging to class P_1 and value 1 indicates oil belonging to class P_2.

Therefore, the neuron that implements the Perceptron has three inputs and one output, as illustrated in Fig. 3.8.

Using the supervised Hebb's algorithm (Hebb's rule) for pattern classification and, assuming the learning rate as 0.01, do the following tasks:

1. Execute five training processes for the Perceptron network, initializing the weight vector $\{w\}$ with random values between zero and one for each training processes. If necessary, update the random number generator in each process so that the initial elements composing the vector are different on each training. The training set is found in Appendix A.
2. Record the results from the five training processes on Table 3.2.

Fig. 3.8 Perceptron architecture for the practical work

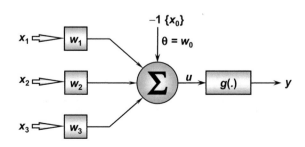

Table 3.2 Results from the Perceptron training

Training	Vector of weights (initial)				Vector of weights (final)				Number of epochs
	w_0	w_1	w_2	w_3	w_0	w_1	w_2	w_3	
#1 (T1)									
#2 (T2)									
#3 (T3)									
#4 (T4)									
#5 (T5)									

Table 3.3 Oil samples for validating the Perceptron network

Sample	x_1	x_2	x_3	y (T1)	y (T2)	y (T3)	y (T4)	y (T5)
1	−0.3665	0.0620	5.9891					
2	−0.7842	1.1267	5.5912					
3	0.3012	0.5611	5.8234					
4	0.7757	1.0648	8.0677					
5	0.1570	0.8028	6.3040					
6	−0.7014	1.0316	3.6005					
7	0.3748	0.1536	6.1537					
8	−0.6920	0.9404	4.4058					
9	−1.3970	0.7141	4.9263					
10	−1.8842	−0.2805	1.2548					

3. After training the Perceptron put the network into operation to classify the oil samples from Table 3.3, indicating on this table the output values (classes) from the five training processes performed on item 1.

4. Explain why the number of training epochs of this application varies each time the Perceptron is trained.

5. For this given application, is it possible to affirm that the classes are linearly separable?

Chapter 4
The ADALINE Network and Delta Rule

4.1 Introduction

The ADALINE (Adaptive Linear Element) was created by Widrow and Hoff in 1960. Its main application was in switching circuits of telephone networks, which was one of the first industrial applications that effectively involved artificial neural networks (Widrow and Hoff 1960).

Despite being a simple network, the ADALINE promoted some essential advancements to the artificial neural network area. Among these contributions, it is possible to mention the following:

- Development of the "Delta rule" learning algorithm.
- Its application in several practical problems involving analog signal processing.
- The first industrial applications of artificial neural networks.

More specifically, the major contribution of the ADALINE was the introduction of a learning algorithm named Delta rule, which is considered the precursor of the generalized Delta rule used on the training of the multiple-layer Perceptron (which will be studied in Chap. 5).

In a similar fashion to the structural configuration of the Perceptron, the ADALINE also consists of a single neural layer, and a single artificial neuron composes it. The arrangement of several ADALINEs into a single network is called MADALINE (Multiple ADALINE) (Widrow and Winter 1988).

Figure 4.1 illustrates an ADALINE network composed of n input signals and only one output. Notice that the same is composed of a single neuron.

According to the classification presented in Chap. 2, the ADALINE network also belongs to the single-layer feedforward architecture, because the information flows forward, that is, from the inputs to the output. It is possible to verify that there is no feedback of values produced by its single neuron.

© Springer International Publishing Switzerland 2017
I.N. da Silva et al., *Artificial Neural Networks*,
DOI 10.1007/978-3-319-43162-8_4

Just like the Perceptron, due to its structural simplicity, the ADALINE network is majorly used on pattern classification problems involving only two distinct classes.

4.2   Operating Principle of the ADALINE

From the analysis of Fig. 4.1, it is possible to see that each input $\{x_i\}$, which represents the signals from a given application in the external environment, is initially weighted by its respective synaptic weights, which are adjusted during the ADALINE training process.

In sequence, similar to the Perceptron, the ADALINE computes the activation potential $\{u\}$ by summing up all the contributions from the multiplication of the inputs $\{x_i\}$ by the weights $\{w_i\}$, and its threshold $\{\theta\}$. The last step for producing the ADALINE output $\{y\}$ is using of an activation function $g(u)$, which usually consists of the step (1.3) or bipolar step (1.5) function.

Thus, the steps required to obtain the ADALINE output $\{y\}$ use the same sequence defined for the Perceptron. Such computation is given by the following expressions:

$$u = \sum_{i=1}^{n} w_i \cdot x_i - \theta \Leftrightarrow u = \sum_{i=0}^{n} w_i \cdot x_i \qquad (4.1)$$

$$y = g(u), \qquad (4.2)$$

where x_i are the input signals of the ADALINE, w_i is the synaptic weight associated to the ith input, θ is the activation threshold or bias, $g(.)$ is the activation function and u is the activation potential.

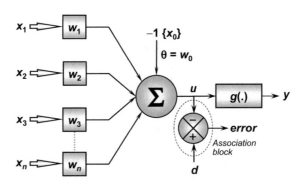

Fig. 4.1 Illustration of the ADALINE network

Table 4.1 Aspects of the ADALINE parameters

Parameter	Variable	Type
Inputs	x_i (ith input)	Real or binary (from the external environment)
Synaptic weights	w_i (associated with x_i)	Real (initialized with random values)
Threshold (bias)	θ	Real (initialized with random values)
Output	y	Binary
Activation function	$g(.)$	Step or bipolar step function
Training process	–	Supervised
Learning rule	–	Delta rule

Since the ADALINE is still mostly used in applications involving pattern recognition, its outputs can classify only two classes, which are associated to one of the two possible values produced by the adopted activation function (step or bipolar step).

Still regarding Fig. 4.1, it is possible to observe the existence of an association block on the ADALINE structure, whose function is to help with the training process of the network, which is detailed in next section. The error signal produced by this block is given by:

$$\text{error} = d - u \qquad (4.3)$$

To summarize, considering a single training sample, the value of the difference (error) between the activation potential $\{u\}$ produced by the network and its respective desired value $\{d\}$ is used to adjust the weights $\{w_0, w_1, w_2, \ldots, w_n\}$ of the network.

Table 4.1 shows the characteristics of the parameters related to the operation dynamics of the ADALINE.

From Table 4.1, it is possible to see that one of the main differences between the ADALINE and the Perceptron is on the learning rule used to adjust weights and threshold.

The same mathematical analysis performed to demonstrate the convergence conditions for the Perceptron is also applied to the ADALINE. In summary, the classes from the problem being mapped must be linearly separable in order to be completely identified by the network.

4.3   Training Process of the ADALINE

The process of adjusting the weights and threshold of the ADALINE network is based on a learning algorithm named the Delta rule (Widrow and Hoff 1960) or Widrow-Hoff learning rule, also known as LMS (Least Mean Square) algorithm or Gradient Descent method.

The idea of the Delta rule, when p training samples are available, is to minimize the difference between the desired output $\{d\}$ and the response $\{u\}$ of the linear combiner, considering all p samples, in order to adjust the weights and threshold of the neuron.

More specifically, the rule performs the minimization of the squared error between u and d to adjust the weight vector $w = \begin{bmatrix} \theta & w_1 & w_2 \cdots w_n \end{bmatrix}^T$ of the network. In summary, the objective is to obtain an optimal w^* so that the squared error $\{E(w^*)\}$ of the whole sample set is as low as possible. In mathematical notation, considering an optimal weight configuration, it can be stated that:

$$E(w^*) \leq E(w), \text{ for } \forall w \in \Re^{n+1} \tag{4.4}$$

The function of the squared error related to the p training samples is defined by:

$$E(w) = \frac{1}{2} \sum_{k=1}^{p} (d^{(k)} - u)^2 \tag{4.5}$$

Substituting the result of (4.1) into (4.5) outcomes:

$$E(w) = \frac{1}{2} \sum_{k=1}^{p} \left(d^{(k)} - \left(\sum_{i=1}^{n} w_i \cdot x_i^{(k)} - \theta \right) \right)^2 \tag{4.6}$$

$$E(w) = \frac{1}{2} \sum_{k=1}^{p} \left(d^{(k)} - (w^T \cdot x^{(k)} - \theta) \right)^2 \tag{4.7}$$

Thus, expression (4.7) computes the mean squared error by processing the p training samples provided for the training process of the ADALINE.

The next step consists of the application of the gradient operator on the mean squared error with respect to vector w, in order to search for an optimal value for the squared error function given by (4.7), that is:

$$\nabla E(w) = \frac{\partial E(w)}{\partial w} \tag{4.8}$$

Using the result of (4.7) in (4.8) results:

$$\nabla E(\boldsymbol{w}) = \sum_{k=1}^{p} \left(d^{(k)} - (\boldsymbol{w}^T \cdot \boldsymbol{x}^{(k)} - \theta) \right) \cdot (-\boldsymbol{x}^{(k)}) \qquad (4.9)$$

Returning the value of (4.1) into (4.10) results:

$$\nabla E(\boldsymbol{w}) = -\sum_{k=1}^{p} (d^{(k)} - u) \cdot \boldsymbol{x}^{(k)} \qquad (4.10)$$

Finally, the steps for adapting the weight vector must be executed in the opposite direction of the gradient, because the optimization goal is to minimize the squared error. In this condition, the variation $\Delta \boldsymbol{w}$ to update the ADALINE weight vector is given by:

$$\Delta \boldsymbol{w} = -\eta \cdot \nabla E(\mathbf{w}) \qquad (4.11)$$

Inserting the result of (4.10) in (4.11), we obtain:

$$\Delta \boldsymbol{w} = \eta \sum_{k=1}^{p} (d^{(k)} - u) \cdot \boldsymbol{x}^{(k)} \qquad (4.12)$$

In a complementary way, we can express (4.12) by:

$$w^{\text{current}} = w^{\text{previous}} + \eta \sum_{k=1}^{p} (d^{(k)} - u) \cdot \boldsymbol{x}^{(k)} \qquad (4.13)$$

Alternatively, to simplify the expression, the update of $\boldsymbol{w}$ can be performed discretely after presenting each kth training sample, that is:

$$w^{\text{current}} = w^{\text{previous}} + \eta \cdot (d^{(k)} - u) \cdot \boldsymbol{x}^{(k)}, \quad \text{with } k = 1 \ldots p \qquad (4.14)$$

In algorithmic notation, the previous expression is equivalent to

$$\boldsymbol{w} \leftarrow \boldsymbol{w} + \eta \cdot (d^{(k)} - u) \cdot \boldsymbol{x}^{(k)}, \quad \text{with } k = 1 \ldots p, \qquad (4.15)$$

where:

$\boldsymbol{w} = [\theta \quad w_1 \quad w_2 \cdots w_n]^T$ is the vector containing the threshold and weights;
$\boldsymbol{x}^{(k)} = [-1 \quad x_1^{(k)} \quad x_2^{(k)} \cdots x_n^{(k)}]^T$ is the kth training sample;
$d^{(k)}$ is the kth training sample;
u is the output value of the linear additive element; and
η is a constant that defines the learning rate.

Similarly to the Perceptron, the learning rate η defines how fast the network training process advances in the direction of the minimum point of the squared error function given by (4.5). Usually, it assumes values within the range of $0 < \eta < 1$.

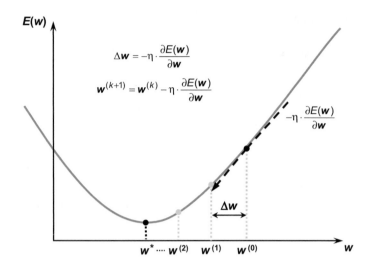

$$\Delta w = -\eta \cdot \frac{\partial E(w)}{\partial w}$$

$$w^{(k+1)} = w^{(k)} - \eta \cdot \frac{\partial E(w)}{\partial w}$$

$$-\eta \cdot \frac{\partial E(w)}{\partial w}$$

Fig. 4.2 Geometric interpretation of the Delta rule

The expression (4.15) is also similar to the one presented on (3.8) because, for the Perceptron, the result produced by the output $\{y\}$ of its neuron is used on the adjustment rule for its internal parameters.

To demonstrate the convergence process of the ADALINE, Fig. 4.2 illustrates a geometric interpretation of the steps needed to update vector w towards the minimum point w^* of the squared error function $\{E(w)\}$.

Hence, from Fig. 4.2 it is possible to observe that from an initial vector $\{w^{(0)}\}$, the next value of w (represented by $w^{(1)}$) is obtained by considering the opposite direction of the gradient vector with respect to $w^{(0)}$. For the next updating step, the adjustment of w (now represented by $w^{(2)}$) is performed by considering the value of the gradient with respect to $w^{(1)}$.

By successively employing such steps, the convergence process will advance iteratively towards w^*, which is the optimal configuration of the internal parameters of the ADALINE. After the process has converged for w^*, the value of $E(w^*)$ will always be smaller than any $E(w)$ calculated in the previous steps.

The step-by-step sequence for training the ADALINE, considering the same matrix form used on training the Perceptron (Sect. 3.4), is expressed using the pseudocode as follows. The stopping criterion is stipulated by employing the mean squared error $\{\bar{E}(w)\}$ with respect to all training samples, and is defined by:

$$\bar{E}(w) = \frac{1}{p}\sum_{k=1}^{p}\left(d^{(k)} - u\right)^2 \tag{4.16}$$

The algorithm converges when the difference of the mean squared error between two successive epochs is small enough, that is:

$$\left|\bar{E}(w^{\text{current}}) - \bar{E}(w^{\text{previous}})\right| \le \varepsilon, \tag{4.17}$$

where ε is the precision required for the convergence process. It is specified by taking into account the performance requirements of the application being mapped by the ADALINE network.

Begin {ADALINE Algorithm – Training Phase}

<1> Obtain the training sample set $\{x^{(k)}\}$;

<2> Associate the desired output $\{d^{(k)}\}$ for each obtained sample;

<3> Initialize the vector w with small random values;

<4> Specify the learning rate $\{\eta\}$ and the required precision $\{\varepsilon\}$;

<5> Initialize the epoch counter $\{epoch \leftarrow 0\}$;

<6> Repeat:

  <6.1> $\bar{E}^{previous} \leftarrow \bar{E}(w)$;

  <6.2> For all training samples $\{x^{(k)}, d^{(k)}\}$, do:

    <6.2.1> $u \leftarrow w^T \cdot x^{(k)}$;

    <6.2.2> $w \leftarrow w + \eta \cdot (d^{(k)} - u) \cdot x^{(k)}$;

  <6.3> $epoch \leftarrow epoch + 1$;

  <6.4> $\bar{E}^{current} \leftarrow \bar{E}(w)$;

  Until: $\left|\bar{E}^{current} - \bar{E}^{previous}\right| \le \varepsilon$

End {ADALINE Algorithm – Training Phase}

As already seen on the Perceptron training, it is possible to verify that the variable *epoch* is responsible for counting the number of training epochs, that is, how many times it is necessary to present all samples of the training set to adjust the weight vector $\{w\}$. The ADALINE network will be considered trained when the difference of the mean squared error between two successive epochs is smaller than the precision $\{\varepsilon\}$ required for the problem being mapped by the ADALINE. The algorithm used to obtain the Mean Squared Error (MSE) given by (4.16) can be implemented as follows

Begin {MSE Algorithm}

<1> Obtain the number of training samples {p};

<2> Initialize variable $\bar{E}$ with zero value {$\bar{E} \leftarrow 0$};

<3> For all training inputs {$\mathbf{x}^{(k)}$, $d^{(k)}$}, do:

$\qquad$ <3.1> $u \leftarrow \mathbf{w}^T \cdot \mathbf{x}^{(k)}$;

$\qquad$ <3.1> $\bar{E} \leftarrow \bar{E} + (d^{(k)} - u)^2$;

<4> $\bar{E} \leftarrow \dfrac{\bar{E}}{p}$;

Begin {MSE Algorithm}

After the training process ends, the ADALINE is then ready to a pattern classification task when new samples are presented to its inputs. The algorithmic steps for this operating phase are given as follow.

Begin {ADALINE Algorithm – Operation Phase}

<1> Obtain a sample to be classified {$\mathbf{x}$};

<2> Use vector $\mathbf{w}$ that was adjusted during training;

<3> Execute the following instructions:

$\qquad$ <3.1> $u \leftarrow \mathbf{w}^T \cdot \mathbf{x}$;

$\qquad$ <3.2> $y \leftarrow \text{signal}(u)$;

$\qquad$ <3.3> If $y = -1$

$\qquad\qquad$ <3.3.1> then: sample $\mathbf{x} \in$ {*Class A*}

$\qquad$ <3.4> If $y = 1$

$\qquad\qquad$ <3.4.1> then: sample $\mathbf{x} \in$ {*Class B*}

Begin {ADALINE Algorithm – Operation Phase}

Thus, as mentioned before, the ADALINE training process tends to move the weight vector, continuously, to minimize the squared error related to all the samples available in the learning process.

To illustrate how the ADALINE training process advances toward the decision boundary between two classes, Fig. 4.3 presents two scenarios showing this convergence. Here, only two inputs {x_1 and x_2} were considered, for teaching purposes.

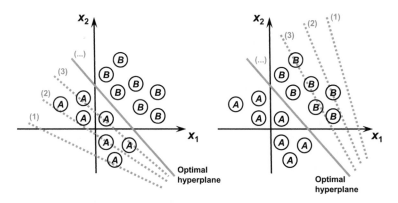

Fig. 4.3 Illustration of the ADALINE training process

From Fig. 4.3, it is possible to notice that for both cases, where the initial weight vector $w^{(0)}$ started in different positions, the convergence process moves the hyperplane toward the optimal separability boundary, which corresponds to the w^* value that minimizes the squared error function (Fig. 4.2).

Figure 4.4 presents the behavior of the mean squared error ($\bar{E}$) with respect to the number of training epochs.

Thus, Fig. 4.4 shows that the mean squared error curve of the ADALINE is always descendant, meaning that it decreases as the number of training epochs increases. It establishes at a constant value when the minimum point of the squared error function (Fig. 4.2) is reached.

Fig. 4.4 Behavior of the mean squared error function with respect to the training epochs

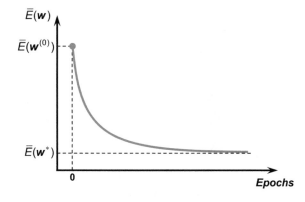

4.4   Comparison Between the Training Processes of the Perceptron and the ADALINE

As discussed in the previous section, the ADALINE training process is performed by employing the Delta rule, whose objective is to minimize the difference between the desired output $\{d\}$ and the response $\{u\}$ of the linear additive element, taking into account all the available training samples. In this case, independently from the initial values attributed to its weight vector, the separation hyperplane obtained after the convergence of the network is always the same.

Regarding the Perceptron training process, the steps taken to adjust its weights are performed by the use of the Hebb's rule, which considers the response produced after presenting each training sample (local synapses) (Sect. 3.4). In such scenario, any hyperplane positioned within the separability area of both classes will be considered an appropriated solution for purposes of pattern classification for the Perceptron.

Figure 4.5 presents two illustrations, which synthesize the discussion of both previous paragraphs.

From Fig. 4.5, it is possible to verify that for the Perceptron, the hyperplane that separates classes can have infinite dispositions because the final configuration of the weight vector is strongly dependent on the initial values randomly attributed to it.

For the ADALINE, the hyperplane parameters are adjusted by applying the Least Mean Square (LMS) method, which is the essence of the Delta rule. Consequently, independently from the initial values attributed to its weight vector, the final configuration of the hyperplane will always be the same. Such procedure makes the ADALINE a neural network with better immunity to eventual noises affecting the process it is mapping.

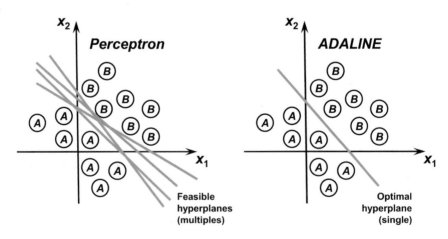

Fig. 4.5 Comparison between the separation of classes performed by the Perceptron and the ADALINE

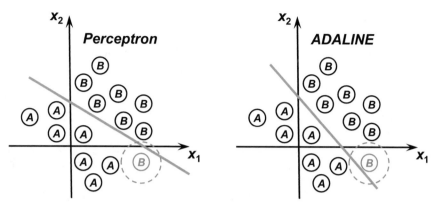

Fig. 4.6 Application of the Perceptron and the ADALINE when considering a noisy sample

To illustrate this characteristic, Fig. 4.6 presents a scenario where both networks classify a new noisy sample. On this scenario, the sample might be affected by some additive noise within the area represented by the dashed circle.

From the analysis of Fig. 4.6, where a noisy condition is considered, the Perceptron network presents a higher probability of incorrectly classifying the given sample, while the ADALINE network has better chances of classifying it correctly, since its hyperplane is positioned on the optimal decision boundary considering the "Least Mean Square" error algorithm.

Hence, considering the several aspects involved in the ADALINE training process, some practical notes on its convergence are presented next.

(a) Differently from the Perceptron, even when the classes of the problem being mapped are nonlinearly separable, the Delta rule is liable to convergence due to the level of precision used to measure the difference between the mean squared errors from two successive epochs (Reed and Marks II 1999).

(b) The value of the learning rate $\{\eta\}$ must be carefully specified to avoid instability around the minimum point of the mean squared error function, as well to prevent the convergence process from becoming excessively slow.

(c) Like in the Perceptron, the number of epochs required for the ADALINE convergence varies according to the initial values attributed to its weight vector $\{w\}$, to the spatial disposition of training samples, and to the value used for the learning rate $\{\eta\}$.

(d) The final position of the hyperplane obtained after convergence provides to the ADALINE better robustness with respect to an eventual noise that may corrupt the samples being classified.

(e) The performance of the ADALINE training can be improved by the normalization of input signals to appropriated domains, as it is detailed in next chapter.

4.5   Exercises

1. Consider a problem to be mapped by the ADALINE with nonlinearly separable classes. Explain if the training process (using the Delta rule) will converge for the given scenario.
2. Explain why the ADALINE training is processed faster than the Perceptron training. Consider that both networks are applied to the same problem with the same initial weight vector.
3. From the previous exercise, discuss an eventual strategy that uses an already-trained ADALINE network to verify whether a given problem is linearly separable or not.
4. Explain the main differences between the Perceptron and the ADALINE.
5. Consider the application of artificial neural networks in a pattern classification problem that needs on-line training. Explain which kind of network (Perceptron or ADALINE) would be more appropriated.
6. Based on the training process using the Delta rule, explain the possible instabilities that could be verified when higher values are adopted for the learning rate. Explain if there are possible inconveniences when extremely lower values are used for the learning rate.
7. Considering the steps for deriving the ADALINE learning process, explain if it is possible to use the neuron output $\{y\}$ in expression (4.5) instead of the activation potential $\{u\}$.
8. Discuss the following statement and indicate whether it is true or false: independently from the initial values of the weight vector of the ADALINE, the same final configuration of w^* will be reached after its convergence.
9. Considering the previous question, explain if the number of training epochs will also be the same, independently of the initial values of the weight vector.
10. Regarding the stop criterion for the convergence process of the ADALINE, given by (4.17), explain if it is necessary to apply the modulus operator on the difference between the mean squared errors of two successive epochs.

4.6   Practical Work

A system for the automatic management of two valves, placed 500 m (around 1600 feet) away of an industrial process, sends a codified signal with four variables $\{x_1, x_2, x_3$ and $x_4\}$ that are necessary for the operation of the valve actuator. As illustrated in Fig. 4.7, the same communication line is used for activating both valves, so the switch placed next to them must decide whether the signal is for valve A or B.

However, during the communication, the signals suffer interferences that modify the content of the originally transmitted information. To avoid this problem, the team of engineers and scientists decided to train an ADALINE to classify the noisy

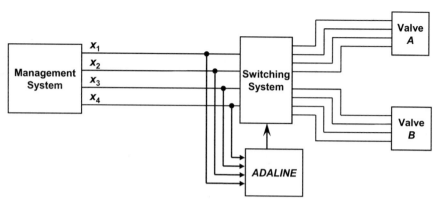

Fig. 4.7 Schematic structure of the valve management system

signals, in order to ensure to the switching system what data must be forwarded to valve *A* or *B*.

Thus, using the measurement of some noisy samples, the team compiled the training set presented in Appendix B, with the following convention: value −1 for the signals that must be sent to valve *A*; and value 1 for the signals that must be sent to valve *B*. The ADALINE structure proposed to this task is illustrated in Fig. 4.8.

Using the learning algorithm Delta rule for pattern classification with the ADALINE, perform the following activities:

1. Execute 5 training processes for the ADALINE, initializing the weight vector $\{w\}$ with random values between zero and one for each training process. If necessary, update the random number generator in each training so that the initial elements composing the vector are different in each process. Use a learning rate value $\{\eta\}$ equal to 0.0025 and a precision $\{\varepsilon\}$ equal to 10^{-6}. The training set can be found in Appendix B.
2. Register the results from the 5 training processes on Table 4.2

Fig. 4.8 Topology of the ADALINE used in the practical work

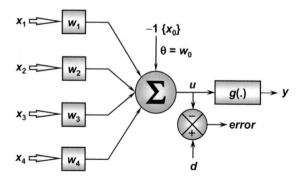

Table 4.2 Results from the ADALINE training

Training	Vector of weights (initial)					Vector of weights (final)					Number of epochs
	w_0	w_1	w_2	w_3	w_4	w_0	w_1	w_2	w_3	w_4	
#1 (T1)											
#2 (T2)											
#3 (T3)											
#4 (T4)											
#5 (T5)											

Table 4.3 Signal samples for classification by the ADALINE

Sample	x_1	x_2	x_3	x_4	y (T1)	y (T2)	y (T3)	y (T4)	y (T5)
1	0.9694	0.6909	0.4334	3.4965					
2	0.5427	1.3832	0.6390	4.0352					
3	0.6081	−0.9196	0.5925	0.1016					
4	−0.1618	0.4694	0.2030	3.0117					
5	0.1870	−0.2578	0.6124	1.7749					
6	0.4891	−0.5276	0.4378	0.6439					
7	0.3777	2.0149	0.7423	3.3932					
8	1.1498	−0.4067	0.2469	1.5866					
9	0.9325	1.0950	1.0359	3.3591					
10	0.5060	1.3317	0.9222	3.7174					
11	0.0497	−2.0656	0.6124	−0.6585					
12	0.4004	3.5369	0.9766	5.3532					
13	−0.1874	1.3343	0.5374	3.2189					
14	0.5060	1.3317	0.9222	3.7174					
15	1.6375	−0.7911	0.7537	0.5515					

3. For the first two training processes, plot a graph showing the mean squared error on each training epoch, and analyze the behavior of both graphs. Discuss if the classes involved with the problem can be considered linearly separable.

4. For all training processes previously executed, present the values registered in Table 4.3 to the ADALINE (already trained), in order to classify (indicate to the switch) if the given signals should be sent to valve A or B.

5. Knowing that although the number of epochs of each training process from item 2 may be different, explain why the final values of the weights remain almost the same.

Chapter 5
Multilayer Perceptron Networks

5.1 Introduction

Multilayer Perceptron (MLP) network features, at least, one intermediate (hidden) neural layer, which is placed between the input layer and the respective output layer. Consequently, MLP networks have at least two neural layers, and their neurons are distributed among the intermediate and output layers.

MLP networks are also known for their wide range of application in several problems from different knowledge areas and are also considered one of the most versatile architectures regarding applicability. Among these potential areas, the most important are the following:

- Universal function approximation (curve fitting).
- Pattern recognition.
- Process identification and control.
- Time series forecasting (prediction).
- System optimization.

According to the classification presented in Chap. 2, MLP networks belong to the multiple layer feedforward architecture, whose training is performed with a supervised process. As shown in Fig. 5.1, the flow of information within the network starts in the input layer, passes through the intermediate layers, and finishes with the output neural layer. Also, conventional MLP networks do not have any feedback of values produced by either the output neural layer or the intermediate neural layers.

The beginning of the great popularity and vast applicability of the MLP networks dates to the late 80s, and such effects are accredited to the publication of the book "Parallel Distributed Processing" (Rumelhart et al. 1986), where the learning algorithm called backpropagation was consistently explained, thus allowing its application on the learning process of these networks.

© Springer International Publishing Switzerland 2017
I.N. da Silva et al., *Artificial Neural Networks*,
DOI 10.1007/978-3-319-43162-8_5

5.2   Operating Principle of the Multilayer Perceptron

By analyzing Fig. 5.1, it is possible to see that each input of the network, which represents the signals from a given application, will be propagated layer-by-layer toward the output layer. In this case, the outputs of the neurons from the first neural layer will be the inputs of the neurons from the second hidden neural layer. For the scenario illustrated in Fig. 5.1, the neurons of the second hidden neural layer will be the inputs to the neurons of the output neural layer.

Thus, the propagation of the input signals of an MLP, independently of the number of intermediate layers, always flows in one direction, that is, from the input layer to the output neural layer.

Differently from the Perceptron and ADALINE networks, it is possible to see in Fig. 5.1, besides the existence of hidden layers on the MLP topology that the output layer can be composed of several neurons, and each of these neurons would represent one of the outputs of the process being mapped. Thus, if such process consists of m outputs, then the MLP network would also have m neurons on its last neural layer.

In summary, in contrast to the Perceptron and ADALINE networks, in which a single neuron is responsible for the full mapping of the whole process, the knowledge related to the behavior of the input and outputs of the system will be distributed among all neurons composing the MLP. The stimuli or signals are presented to the network on its input layer. The intermediate layers, on the other hand, extract the majority of the information related to the system behavior and codify them using the synaptic weights and thresholds of their neurons, thus forming a representation of the environment where the particular system exists. Finally, the neurons of the output layer receive the stimuli from the neurons of the

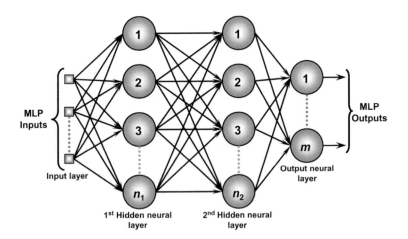

Fig. 5.1 Illustration of a multilayer perceptron network

last intermediate layer, producing a response pattern which will be the output generated by the network.

The specification of the topological configuration of an MLP network, as well as the number of intermediate layers and their respective number of neurons, depends on several factors that will be addressed in this chapter. More specifically, the class of the problem being mapped by the MLP, the spatial distribution of the training samples, and the initial values for both the training parameters and weight matrices are elements which help to set up the topology of the network.

As mentioned, the adjustment of the weights and thresholds of each neuron of an MLP network is made by employing a supervised training process, that is, for each sample of input data, there must be the respective desired output (response).

The learning algorithm used during the training process of an MLP is called backpropagation.

5.3   Training Process of the Multilayer Perceptron

The training process of MLP networks using the backpropagation algorithm, also known as the generalized Delta rule, is usually done by the successive application of two specific stages. These stages are illustrated in Fig. 5.2, which shows an MLP configuration composed of two hidden layers, n signals on its input layer, n_1 neurons in its first hidden neural layer, n_2 neurons in its second hidden neural layer, and n_3 signals associated with the output neural layer (third neural layer).

The first stage is called forward propagation, where the signals $\{x_1, x_2, ..., x_n\}$ of a given sample from the training set are inserted into the network inputs and are propagated layer-by-layer until the production of the corresponding outputs. Thus,

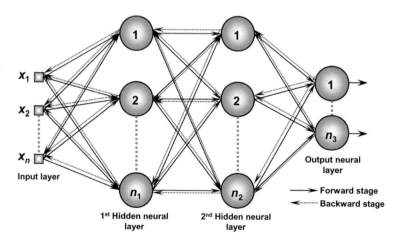

Fig. 5.2 Illustration of both training stages of the MLP network

this stage intends solely in obtaining the responses from the network, taking into account only the current values of the synaptic weights and thresholds of its neurons, which will remain unmodified during the execution of this stage.

Next, the responses produced by the network outputs are compared to the respective available desired responses, since it is a supervised learning process, as mentioned earlier. It is important to note that, considering an MLP network with n_3 neurons in its output layer, the respective n_3 deviations (errors) between the desired responses and those produced by the output neurons are calculated (Sect. 5.3.1) and will be used after that to adjust the weights and thresholds of all neurons.

Therefore, because of these errors, it is applied the second stage of the backpropagation algorithm, known as backward propagation. Unlike the first stage, the modifications (adjustments) of the synaptic weights and thresholds of all neurons of the network are executed during this stage.

In summary, the successive application of the forward and backward stages allows the synaptic weights and thresholds of the neurons to be adjusted automatically in each iteration, also resulting in the gradual reduction of the sum of the errors produced by the network responses with respect to the desired responses.

5.3.1   Deriving the Backpropagation Algorithm

For a better understanding of the working principle of the backpropagation algorithm, it is necessary to define, a priori, several variables and auxiliary parameters that will be used for such purpose. Based on the MLP topology illustrated in Figs. 5.2 and 5.3 presents the set of variables that will guide the algorithm derivation.

Each neuron (j) belonging to one of the layers (L) from the topology illustrated in Fig. 5.3 can be configured using the topology adopted in Fig. 5.4, where $g(.)$ represents an activation function which must be continuous and differentiable in all its domain, such as those represented by logistic or hyperbolic tangent activation functions.

From Figs. 5.3 and 5.4, the following terminology will be assumed for its fundamental parameters:

- $W_{ji}^{(L)}$ are weight matrices whose elements denote the value of the synaptic weight that connects the jth neuron of layer (L) to the ith neuron of layer ($L - 1$). For the topology shown in Fig. 5.3, we have:

  - $W_{ji}^{(3)}$ is the synaptic weight connecting the jth neuron of output layer to the ith neuron of layer 2.
  - $W_{ji}^{(2)}$ is the synaptic weight connecting the jth neuron of hidden layer 2 to the ith neuron of hidden layer 1.

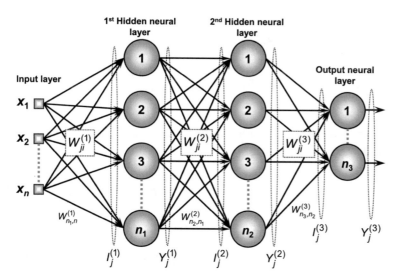

Fig. 5.3 Notation for deriving the backpropagation algorithm

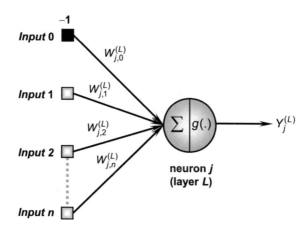

Fig. 5.4 Configuration of a
neuron used for deriving the
backpropagation algorithm

– $W_{ji}^{(1)}$ is the synaptic weight connecting the jth neuron of hidden layer 1 to the
ith signal of the input layer.

• $I_j^{(L)}$ are vectors whose elements denote the weighted inputs related to the jth
neuron of layer L, and are defined by:

$$I_j^{(1)} = \sum_{i=0}^{n} W_{ji}^{(1)} \cdot x_i \Leftrightarrow I_j^{(1)} = W_{1,0}^{(1)} \cdot x_0 + W_{1,1}^{(1)} \cdot x_1 + \cdots + W_{1,n}^{(1)} \cdot x_n \qquad (5.1)$$

$$I_j^{(2)} = \sum_{i=0}^{n_1} W_{ji}^{(2)} \cdot Y_i^{(1)} \Leftrightarrow I_j^{(2)} = W_{1,0}^{(2)} \cdot Y_0^{(1)} + W_{1,1}^{(2)} \cdot Y_1^{(1)} + \cdots + W_{1,n_1}^{(2)} \cdot Y_{n_1}^{(1)} \quad (5.2)$$

$$I_j^{(3)} = \sum_{i=0}^{n_2} W_{ji}^{(3)} \cdot Y_i^{(2)} \Leftrightarrow I_j^{(3)} = W_{1,0}^{(3)} \cdot Y_0^{(2)} + W_{1,1}^{(3)} \cdot Y_1^{(2)} + \cdots + W_{1,n_2}^{(3)} \cdot Y_{n_2}^{(2)} \quad (5.3)$$

- $Y_j^{(L)}$ are vectors whose elements denote the output of the jth neuron related to the layer L. They are defined as:

$$Y_j^{(1)} = g\left(I_j^{(1)}\right) \quad (5.4)$$

$$Y_j^{(2)} = g\left(I_j^{(2)}\right) \quad (5.5)$$

$$Y_j^{(3)} = g\left(I_j^{(3)}\right) \quad (5.6)$$

To exemplify the adopted terminology, the MLP in Fig. 5.5 is considered. This MLP is composed of two inputs x_1 and x_2 ($n = 2$), three neurons in the first hidden layer ($n_1 = 3$), two neurons in the second hidden layer ($n_2 = 2$) and one output neuron ($n_3 = 1$). For this example, the hyperbolic tangent activation function will be used for all neurons.

In this case, the configuration of the weight matrices, considering the values in Fig. 5.5, would be defined by:

$$W_{ji}^{(1)} = \begin{bmatrix} 0.2 & 0.4 & 0.5 \\ 0.3 & 0.6 & 0.7 \\ 0.4 & 0.8 & 0.3 \end{bmatrix} ; W_{ji}^{(2)} = \begin{bmatrix} -0.7 & 0.6 & 0.2 & 0.7 \\ -0.3 & 0.7 & 0.2 & 0.8 \end{bmatrix} ;$$

$$W_{ji}^{(3)} = \begin{bmatrix} 0.1 & 0.8 & 0.5 \end{bmatrix}$$

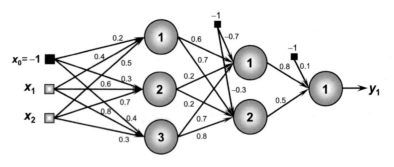

Fig. 5.5 Example of a multilayer perceptron

Assuming an input value defined by $x_1 = 0.3$ and $x_2 = 0.7$, the values of $I_j^{(1)}$ and $Y_j^{(1)}$ would be given by:

$$I_j^{(1)} = \begin{bmatrix} I_1^{(1)} \\ I_2^{(1)} \\ I_3^{(1)} \end{bmatrix} = \begin{bmatrix} W_{1,0}^{(1)} \cdot x_0 + W_{1,1}^{(1)} \cdot x_1 + W_{1,2}^{(1)} \cdot x_2 \\ W_{2,0}^{(1)} \cdot x_0 + W_{2,1}^{(1)} \cdot x_1 + W_{2,2}^{(1)} \cdot x_2 \\ W_{3,0}^{(1)} \cdot x_0 + W_{3,1}^{(1)} \cdot x_1 + W_{3,2}^{(1)} \cdot x_2 \end{bmatrix} = \begin{bmatrix} 0.2 \cdot (-1) + 0.4 \cdot 0.3 + 0.5 \cdot 0.7 \\ 0.3 \cdot (-1) + 0.6 \cdot 0.3 + 0.7 \cdot 0.7 \\ 0.4 \cdot (-1) + 0.8 \cdot 0.3 + 0.3 \cdot 0.7 \end{bmatrix} = \begin{bmatrix} 0.27 \\ 0.37 \\ 0.05 \end{bmatrix}$$

$$Y_j^{(1)} = \begin{bmatrix} Y_1^{(1)} \\ Y_2^{(1)} \\ Y_3^{(1)} \end{bmatrix} = \begin{bmatrix} g\left(I_1^{(1)}\right) \\ g\left(I_2^{(1)}\right) \\ g\left(I_3^{(1)}\right) \end{bmatrix} = \begin{bmatrix} \tanh(0.27) \\ \tanh(0.37) \\ \tanh(0.05) \end{bmatrix} = \begin{bmatrix} 0.26 \\ 0.35 \\ 0.05 \end{bmatrix} \xrightarrow{Y_0^{(1)} = -1} Y_j^{(1)} = \begin{bmatrix} Y_0^{(1)} \\ Y_1^{(1)} \\ Y_2^{(1)} \\ Y_3^{(1)} \end{bmatrix} = \begin{bmatrix} -1 \\ 0.26 \\ 0.35 \\ 0.05 \end{bmatrix},$$

where the arguments used for the hyperbolic tangent function (tanh) are in radians. The vectors $I_j^{(2)}$ and $Y_j^{(2)}$ referring to the second neural layer would be given by:

$$I_j^{(2)} = \begin{bmatrix} I_1^{(2)} \\ I_2^{(2)} \end{bmatrix} = \begin{bmatrix} W_{1,0}^{(2)} \cdot Y_0^{(1)} + W_{1,1}^{(2)} \cdot Y_1^{(1)} + W_{1,2}^{(2)} \cdot Y_2^{(1)} + W_{1,3}^{(2)} \cdot Y_3^{(1)} \\ W_{2,0}^{(2)} \cdot Y_0^{(1)} + W_{2,1}^{(2)} \cdot Y_1^{(1)} + W_{2,2}^{(2)} \cdot Y_2^{(1)} + W_{2,3}^{(2)} \cdot Y_3^{(1)} \end{bmatrix} = \begin{bmatrix} 0.96 \\ 0.59 \end{bmatrix}$$

$$Y_j^{(2)} = \begin{bmatrix} Y_1^{(2)} \\ Y_2^{(2)} \end{bmatrix} = \begin{bmatrix} g(I_1^{(2)}) \\ g(I_2^{(2)}) \end{bmatrix} = \begin{bmatrix} \tanh(0.96) \\ \tanh(0.59) \end{bmatrix} = \begin{bmatrix} 0.74 \\ 0.53 \end{bmatrix} \xrightarrow{Y_0^{(2)} = -1} Y_j^{(2)} = \begin{bmatrix} Y_0^{(2)} \\ Y_1^{(2)} \\ Y_2^{(2)} \end{bmatrix} = \begin{bmatrix} -1 \\ 0.74 \\ 0.53 \end{bmatrix}$$

Finally, the vectors $I_j^{(3)}$ and $Y_j^{(3)}$ of the third neural layer would be given by:

$$I_j^{(3)} = \begin{bmatrix} I_1^{(3)} \end{bmatrix} = \begin{bmatrix} W_{1,0}^{(3)} \cdot Y_0^{(2)} + W_{1,1}^{(3)} \cdot Y_1^{(2)} + W_{1,2}^{(3)} \cdot Y_2^{(2)} \end{bmatrix} = [0.76]$$

$$Y_j^{(3)} = \begin{bmatrix} Y_1^{(3)} \end{bmatrix} = \begin{bmatrix} g(I_1^{(3)}) \end{bmatrix} = [\tanh(0.76)] = [0.64]$$

In this last expression, there is no need to insert the variable $Y_0^{(3)} = -1$, since this is the last neural layer, and the value of $Y_1^{(3)}$ is the actual output produced by this network.

The next step in the derivation of the backpropagation algorithm consists of defining a function that represents the approximation error, whose purpose is to measure the deviation of the responses produced by the output neurons of the network with respect to the corresponding desired values. Thus, considering the kth training sample for the topology illustrated in Fig. 5.3, the squared error function is employed to measure the local performance associated with the results produced by the output neurons with respect to the given sample, that is:

$$E(k) = \frac{1}{2} \sum_{j=1}^{n_3} \left(d_j(k) - Y_j^{(3)}(k) \right)^2, \tag{5.7}$$

where $Y_j^{(3)}(k)$ is the value produced by the jth output neuron of the network for the kth training sample, while $d_j(k)$ is the corresponding desired value.

Consequently, assuming a training set composed of p samples, the measurement of the global performance of the backpropagation algorithm can be calculated through the "mean squared error" defined by:

$$E_M = \frac{1}{p} \sum_{k=1}^{p} E(k), \tag{5.8}$$

where $E(k)$ is he squared error obtained in (5.7).

To provide a better understanding of the necessary steps for comprehending the backpropagation algorithm, its description will be divided into two parts. The first part is concerned with the adjustment of the synaptic weight matrix $W_{ji}^{(3)}$, representing the output neural layer. The second part concerns the procedures for adjusting weight matrices associated with the intermediate layers that, for the topology illustrated in Fig. 5.3, are composed of $W_{ji}^{(2)}$ and $W_{ji}^{(1)}$.

In this book, the batch or offline learning process (Sect. 2.3.4) will be used to implement the expression (5.8). It will also be used the method based on the gradient of the squared error function given by (5.7).

Part I—Adjusting the Synaptic Weights of the Output Layer

The objective of the training process of the output neural layer consists of adjusting the weight matrix $W_{ji}^{(3)}$ in order to minimize the error between the outputs produced by the network with respect to the desired outputs. In this case, considering the error given by (5.7) related to the jth neuron of the output layer with respect to the kth training sample, the adjustment rule becomes similar to the one applied to the ADALINE network. Thus, by employing the definition of gradient and exploring the chain rule of differentiation of functions, we have:

$$\nabla E^{(3)} = \frac{\partial E}{\partial W_{ji}^{(3)}} = \frac{\partial E}{\partial Y_j^{(3)}} \cdot \frac{\partial Y_j^{(3)}}{\partial I_j^{(3)}} \cdot \frac{\partial I_j^{(3)}}{\partial W_{ji}^{(3)}} \tag{5.9}$$

From the previous definitions, it is obtained:

$$\frac{\partial I_j^{(3)}}{\partial W_{ji}^{(3)}} = Y_i^{(2)} \quad \{\text{Obtained from (5.3)}\} \tag{5.10}$$

$$\frac{\partial Y_j^{(3)}}{\partial I_j^{(3)}} = g'\left(I_j^{(3)}\right) \quad \{\text{Obtained from (5.6)}\} \tag{5.11}$$

$$\frac{\partial E}{\partial Y_j^{(3)}} = -\left(d_j - Y_j^{(3)}\right) \quad \{\text{Obtained from (5.7)}\}, \tag{5.12}$$

where $g'(.)$ denotes the first-order derivative of the employed activation function. Replacing (5.10), (5.11) and (5.12) in (5.9), we have:

$$\frac{\partial E}{\partial W_{ji}^{(3)}} = -\left(d_j - Y_j^{(3)}\right) \cdot g'\left(I_j^{(3)}\right) \cdot Y_i^{(2)} \tag{5.13}$$

Thus, the adjustment of the weight matrix $W_{ji}^{(3)}$ must be made in the opposite direction of the gradient in order to minimize the error, therefore:

$$\Delta W_{ji}^{(3)} = -\eta \cdot \frac{\partial E}{\partial W_{ji}^{(3)}} \quad \Leftrightarrow \quad \Delta W_{ji}^{(3)} = \eta \cdot \delta_j^{(3)} \cdot Y_i^{(2)}, \tag{5.14}$$

where $\delta_j^{(3)}$ is defined as the local gradient related to the jth neuron in the output layer, and is given by:

$$\delta_j^{(3)} = \left(d_j - Y_j^{(3)}\right) \cdot g'\left(I_j^{(3)}\right) \tag{5.15}$$

Additionally, expression (5.14) can also be converted to the following iterative procedure:

$$W_{ji}^{(3)}(t+1) = W_{ji}^{(3)}(t) + \eta \cdot \delta_j^{(3)} \cdot Y_i^{(2)}, \tag{5.16}$$

where η is the learning rate of the backpropagation algorithm. In algorithmic notation, this expression is equivalent to the following:

$$W_{ji}^{(3)} \leftarrow W_{ji}^{(3)} + \eta \cdot \delta_j^{(3)} \cdot Y_i^{(2)} \tag{5.17}$$

Therefore, expression (5.17) adjusts the neuron weights of the output layer of the network by taking into account the difference observed between the responses produced by its outputs and their corresponding desired values.

Part II—Adjusting the Synaptic Weights of the Intermediate Layers

Differently from the neurons belonging to the output layer of the MLP, the neurons of the intermediate layers do not have direct access to the desired values for their outputs. In this scenario, the adjustment of their synaptic weights is performed through estimations of the output errors produced by those neurons of the aftermost layer, which have been already adjusted.

As an example, following the sequence of adjustments obtained by the backward stage for the topology illustrated in Fig. 5.3, the correction of the weights of the neurons of the second intermediate layer is started only after the adjustment of the neurons of the output layer is concluded. In this particular condition, the desired values of the outputs of those neurons are unavailable; thus, their adjustment will be based on the synaptic weights already adjusted for the neurons of the output layer.

Thus, it is in this very aspect that the essence of the backpropagation algorithm dwells, since, first, the adjusted synaptic weights of the neurons of the output layers are calculated by comparing the deviation of the produced responses and the corresponding desired values. Secondly, this error is backpropagated to the neurons of the previous layers, weighted by the values of the synaptic weights that were previously adjusted in all the posterior layers. Consequently, the desired response of a neuron of a hidden layer must be determined with respect to the neurons (from the immediately posterior layer) that are directly connected to it and that have been already adjusted in the previous stage.

(A) Adjusting the synaptic weights of the second hidden layer

The purpose of the training process for the second neural layer consists of adjusting the weight matrix $W_{ji}^{(2)}$ in order to minimize the error between the outputs produced by the network with respect to the backpropagated error originated from the adjustment of the neurons of the output layer. Therefore, we have:

$$\nabla E^{(2)} = \frac{\partial E}{\partial W_{ji}^{(2)}} = \frac{\partial E}{\partial Y_j^{(2)}} \cdot \frac{\partial Y_j^{(2)}}{\partial I_j^{(2)}} \cdot \frac{\partial I_j^{(2)}}{\partial W_{ji}^{(2)}} \qquad (5.18)$$

From the previous definitions, we have:

$$\frac{\partial I_j^{(2)}}{\partial W_{ji}^{(2)}} = Y_i^{(1)} \quad \{\text{Obtained from } (5.2)\} \qquad (5.19)$$

$$\frac{\partial Y_j^{(2)}}{\partial I_j^{(2)}} = g'\left(I_j^{(2)}\right) \quad \{\text{Obtained from } (5.5)\} \qquad (5.20)$$

$$\frac{\partial E}{\partial Y_j^{(2)}} = \sum_{k=1}^{n_3} \frac{\partial E}{\partial I_k^{(3)}} \cdot \frac{\partial I_k^{(3)}}{\partial Y_j^{(2)}} = \underbrace{\sum_{k=1}^{n_3} \frac{\partial E}{\partial I_k^{(3)}}}_{\text{parcel(i)}} \cdot \underbrace{\frac{\partial \left(\sum_{k=1}^{n_3} W_{kj}^{(3)} \cdot Y_j^{(2)}\right)}{\partial Y_j^{(2)}}}_{\text{parcel(ii)}}, \qquad (5.21)$$

where the value of the partial derivative of parcel (ii) with respect to $Y_j^{(2)}$ is the value of $W_{kj}^{(3)}$, that is:

$$\frac{\partial E}{\partial Y_j^{(2)}} = \underbrace{\sum_{k=1}^{n_3} \underbrace{\frac{\partial E}{\partial I_k^{(3)}}}_{\text{parcel(i)}} \cdot \underbrace{W_{kj}^{(3)}}_{\text{parcel(ii)}}} \tag{5.22}$$

Hence, it is important to observe that the value of parcel (ii) in expression (5.22) refers to the synaptic weight of all neurons of the output layer that are interconnected to a given neuron j of the second intermediate layer.

To emphasize the essence of the backpropagation method, notice that all the weights of $W_{ji}^{(3)}$ have been adjusted in the previous step based on real error values, which will be used for adjusting the weights of the second intermediate neural layer. For the sake of comprehension, Fig. 5.6 shows this representation using the neuron j of this layer as a reference point.

Regarding the problem of obtaining the values of parcel (i) of expression (5.22), it is possible to see that its result can be provided by multiplying (5.11) by (5.12), which will result in the actual value of $\delta_j^{(3)}$ obtained in (5.15). Thus, by using such substitutions, expression (5.22) can be represented by:

$$\frac{\partial E}{\partial Y_j^{(2)}} = -\sum_{k=1}^{n_3} \delta_k^{(3)} \cdot W_{kj}^{(3)} \tag{5.23}$$

As consequence, substituting (5.19), (5.20), and (5.23) in (5.18), we have:

$$\frac{\partial E}{\partial W_{ji}^{(2)}} = -\left(\sum_{k=1}^{n_3} \delta_k^{(3)} \cdot W_{kj}^{(3)} \right) \cdot g'\left(I_j^{(2)} \right) \cdot Y_i^{(1)} \tag{5.24}$$

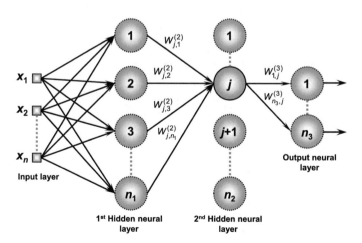

Fig. 5.6 Representation of neural interconnections for adjusting the synaptic weights of the neurons of the second hidden layer

Therefore, the adjustment of the weight matrix $W_{ji}^{(2)}$ must be made in the opposite direction of the gradient in order to minimize the error:

$$\Delta W_{ji}^{(2)} = -\eta \cdot \frac{\partial E}{\partial W_{ji}^{(2)}} \quad \Leftrightarrow \quad \Delta W_{ji}^{(2)} = \eta \cdot \delta_j^{(2)} \cdot Y_i^{(1)}, \qquad (5.25)$$

where $\delta_j^{(2)}$ is defined as the local gradient with respect to the jth neuron of the second intermediate layer, that is:

$$\delta_j^{(2)} = -\left(\sum_{k=1}^{n_3} \delta_k^{(3)} \cdot W_{kj}^{(3)}\right) \cdot g'\left(I_j^{(2)}\right) \qquad (5.26)$$

Complementarily, expression (5.25) can also be converted to the following iterative procedure:

$$W_{ji}^{(2)}(t+1) = W_{ji}^{(2)}(t) + \eta \cdot \delta_j^{(2)} \cdot Y_i^{(1)} \qquad (5.27)$$

In algorithmic notation, this expression is equivalent to:

$$W_{ji}^{(2)} \leftarrow W_{ji}^{(2)} + \eta \cdot \delta_j^{(2)} \cdot Y_i^{(1)} \qquad (5.28)$$

Therefore, expression (5.28) adjusts the weight of the neurons of the second hidden layer, taking into account the backpropagation of the error originated from the neurons of the output layer.

(B) **Adjusting the synaptic weights of the first hidden layer**

The objective of the training process of the first hidden layer consists of adjusting the weight matrix $W_{ji}^{(1)}$ in order to minimize the error between the output produced by the network and the backpropagated error originated by the adjustment of the neurons of the second hidden layer. Thus, we have:

$$\nabla E^{(1)} = \frac{\partial E}{\partial W_{ji}^{(1)}} = \frac{\partial E}{\partial Y_j^{(1)}} \cdot \frac{\partial Y_j^{(1)}}{\partial I_j^{(1)}} \cdot \frac{\partial I_j^{(1)}}{\partial W_{ji}^{(1)}} \qquad (5.29)$$

From the previous definitions, it is obtained:

$$\frac{\partial I_j^{(1)}}{\partial W_{ji}^{(1)}} = x_i \quad \{\text{Obtained from (5.1)}\} \qquad (5.30)$$

$$\frac{\partial Y_j^{(1)}}{\partial I_j^{(1)}} = g'(I_j^{(1)}) \quad \{\text{Obtained from (5.4)}\} \tag{5.31}$$

$$\frac{\partial E}{\partial Y_j^{(1)}} = \sum_{k=1}^{n_2} \frac{\partial E}{\partial I_k^{(2)}} \cdot \frac{\partial I_k^{(2)}}{\partial Y_j^{(1)}} = \sum_{k=1}^{n_2} \underbrace{\frac{\partial E}{\partial I_k^{(2)}}}_{\text{parcel(i)}} \cdot \underbrace{\frac{\partial \left(\sum_{k=1}^{n_2} W_{kj}^{(2)} \cdot Y_j^{(1)} \right)}{\partial Y_j^{(1)}}}_{\text{parcel(ii)}} \tag{5.32}$$

Similarly to (5.22) the value of the partial derivative of the argument of parcel (ii) in (5.32) with respect to $Y_j^{(1)}$ is the very value of $W_{kj}^{(2)}$, which is:

$$\frac{\partial E}{\partial Y_j^{(1)}} = \sum_{k=1}^{n_2} \underbrace{\frac{\partial E}{\partial I_k^{(2)}}}_{\text{parcel(i)}} \cdot \underbrace{W_{kj}^{(2)}}_{\text{parcel(ii)}} \tag{5.33}$$

The same analysis used for adjusting the synaptic weights of the second intermediate neural layer can also be applied for adjusting the weights of the first intermediate layer. In this case, the value of parcel (ii) in (5.33) relates to the synaptic weights of all the neurons of the second intermediate layer that are connected to a given neuron j of the first layer.

It is important to emphasize that all the weights of $W_{ji}^{(2)}$ were adjusted in the previous step based on the backpropagated errors from the adjustment of the weights of $W_{ji}^{(2)}$, which in turn were adjusted based on the real values of the error. Figure 5.7 shows this representation using the neuron j of the first hidden layer as a reference point.

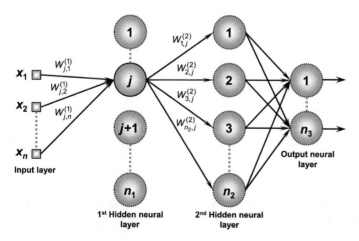

Fig. 5.7 Representation of neural interconnections for adjusting the synaptic weights of the neurons of the first hidden layer

Back to the problem of obtaining the values of the parcel (i) in expression (5.33), it is possible to see that its result can be obtained by multiplying (5.20) by (5.21), which will lead to the value of $\delta_j^{(2)}$ obtained in (5.26). Thus, using such substitutions, expression (5.32) can be represented by:

$$\frac{\partial E}{\partial Y_j^{(1)}} = -\sum_{k=1}^{n_2} \delta_k^{(2)} \cdot W_{kj}^{(2)} \tag{5.34}$$

As consequence, substituting (5.30), (5.31) and (5.34) in (5.29), we have:

$$\frac{\partial E}{\partial W_{ji}^{(1)}} = -\left(\sum_{k=1}^{n_2} \delta_k^{(2)} \cdot W_{kj}^{(2)}\right) \cdot g'\left(I_j^{(1)}\right) \cdot x_i \tag{5.35}$$

Thus, the adjustment of the weight matrix $W_{ji}^{(1)}$ must be made in the opposite direction of the gradient in order to minimize the error:

$$\Delta W_{ji}^{(1)} = -\eta \cdot \frac{\partial E}{\partial W_{ji}^{(1)}} \quad\Leftrightarrow\quad \Delta W_{ji}^{(1)} = \eta \cdot \delta_j^{(1)} \cdot x_i, \tag{5.36}$$

where $\delta_j^{(1)}$ is defined as the local gradient with respect to the jth neuron of the second intermediate layer, that is:

$$\delta_j^{(1)} = -\left(\sum_{k=1}^{n_2} \delta_k^{(2)} \cdot W_{kj}^{(2)}\right) \cdot g'\left(I_j^{(1)}\right) \tag{5.37}$$

Complementarily, expression (5.36) can also be converted to the following iterative procedure:

$$W_{ji}^{(1)}(t+1) = W_{ji}^{(1)}(t) + \eta \cdot \delta_j^{(1)} \cdot x_i \tag{5.38}$$

In algorithmic notation, this expression is equivalent to:

$$W_{ji}^{(1)} \leftarrow W_{ji}^{(1)} + \eta \cdot \delta_j^{(1)} \cdot x_i \tag{5.39}$$

Therefore, expression (5.39) adjusts the weight of the neurons from the second hidden layer, using the backpropagation of the error originated from the neurons of the output layer.

The procedures for adjusting the synaptic weight matrices presented by the previous equations can be generalized for any topology of the MLP network, independently from the number of intermediate layers.

5.3.2   Implementing the Backpropagation Algorithm

Differently from the Perceptron and ADALINE networks, the output layer of the MLP can be constituted by more than one neuron, and such quantity is specified with respect to the number of outputs of the system being mapped.

Thus, in order to clarify the notation used for the vector $x^{(k)}$ and also to its corresponding desired values, it will be assumed, as an example, a problem constituted by three inputs $\{x_1, x_2, x_3\}$ to be mapped by an MLP. The vector x (k) represents the kth training sample and the desired values are inserted in vector $d^{(k)}$. It will also be assumed that the training set is composed of just four samples with the following input values: $\Omega^{(x)} = \{$ [0.2 0.9 0.4]; [0.1 0.3 0.5]; [0.9 0.7 0.8]; [0.6 0.4 0.3] $\}$. In this case, we will also assume that the MLP is solving a curve fitting problem that requires two output variables, which are represented by two neurons belonging to the output layer, whose respective desired values $\{d_1, d_2\}$ for each one of the four inputs are given by: $\Omega^{(d)} = \{$[0.7 0.3]; [0.6 0.4]; [0.9 0.5]; [0.2 0.8]]$\}$. The matrix representation of such values is given by:

$$\Omega^{(x)} = \begin{array}{c} \\ x_0 \\ x_1 \\ x_2 \\ x_3 \end{array} \begin{array}{cccc} x^{(1)} & x^{(2)} & x^{(3)} & x^{(4)} \\ \left[\begin{array}{cccc} -1 & -1 & -1 & -1 \\ 0.2 & 0.1 & 0.9 & 0.6 \\ 0.9 & 0.3 & 0.7 & 0.4 \\ 0.4 & 0.5 & 0.8 & 0.3 \end{array}\right] \end{array} \quad ; \quad \Omega^{(d)} = \begin{array}{c} \\ d_1 \\ d_2 \end{array} \begin{array}{cccc} d^{(1)} & d^{(2)} & d^{(3)} & d^{(4)} \\ \left[\begin{array}{cccc} 0.7 & 0.6 & 0.9 & 0.2 \\ 0.3 & 0.4 & 0.5 & 0.8 \end{array}\right] \end{array}$$

Alternatively, it is possible to extract from both matrices each vector $x^{(k)}$ with the respective vector $d^{(k)}$, which represents each training sample, that is:

$$x^{(1)} = \begin{bmatrix} -1 & 0.2 & 0.9 & 0.4 \end{bmatrix}^T \quad ; \quad \text{with } d^{(1)} = \begin{bmatrix} 0.7 & 0.3 \end{bmatrix}^T$$
$$x^{(2)} = \begin{bmatrix} -1 & 0.1 & 0.3 & 0.5 \end{bmatrix}^T \quad ; \quad \text{with } d^{(1)} = \begin{bmatrix} 0.6 & 0.4 \end{bmatrix}^T$$
$$x^{(3)} = \begin{bmatrix} -1 & 0.9 & 0.7 & 0.8 \end{bmatrix}^T \quad ; \quad \text{with } d^{(1)} = \begin{bmatrix} 0.9 & 0.5 \end{bmatrix}^T$$
$$x^{(4)} = \begin{bmatrix} -1 & 0.6 & 0.4 & 0.3 \end{bmatrix}^T \quad ; \quad \text{with } d^{(1)} = \begin{bmatrix} 0.2 & 0.8 \end{bmatrix}^T$$

The stop criteria of the process is defined by the mean squared error calculated with (5.8), taking into account all training samples available. The algorithm converges when the mean squared error between two successive epochs is sufficiently small, that is:

$$\left| E_M^{\text{current}} - E_M^{\text{previous}} \right| \le \varepsilon, \tag{5.40}$$

where ε is the precision required for the convergence process, and is specified by the type of application that the network is mapping.

The sequence of computational procedures for the training process of the MLP is presented by the following pseudocode:

Begin {MLP Algorithm – Training Phase}

<1> Obtain the set of training samples $\{x^{(k)}\}$;

<2> Associate the vector with the desired output $\{d^{(k)}\}$ for each training sample;

<3> Initialize $w_{ji}^{(1)}$, $w_{ji}^{(2)}$ and $w_{ji}^{(3)}$ with small random values;

<4> Specify the learning rate $\{\eta\}$ and the required precision $\{\varepsilon\}$;

<5> Initialize the epoch counter $\{epoch \leftarrow 0\}$;

<6> Repeat:

　　<6.1> $E_M^{previous} \leftarrow E_M$;　{according to (5.8)}

　　<6.2> For all train samples $\{x^{(k)}, d^{(k)}\}$, do:

　　　　<6.2.1> Obtain $I_j^{(1)}$ and $Y_j^{(1)}$;　{according to (5.1) and (5.4)}

　　　　<6.2.2> Obtain $I_j^{(2)}$ and $Y_j^{(2)}$;　{according to (5.2) and (5.5)}　*Forward steps*

　　　　<6.2.3> Obtain $I_j^{(3)}$ and $Y_j^{(3)}$;　{according to (5.3) and (5.6)}

　　　　<6.2.4> Determine $\delta_j^{(3)}$;　{according to (5.15)}

　　　　<6.2.5> Adjust $w_{ji}^{(3)}$;　{according to (5.17)}

　　　　<6.2.6> Determine $\delta_j^{(2)}$;　{according to (5.26)}　*Backward steps*

　　　　<6.2.7> Adjust $w_{ji}^{(2)}$;　{according to (5.28)}

　　　　<6.2.8> Determine $\delta_j^{(1)}$;　{according to (5.37)}

　　　　<6.2.9> Adjust $w_{ji}^{(1)}$;　{according to (5.39)}

　　<6.3> Obtain the adjusted $y_j^{(3)}$;　{according to <6.2.1>, <6.2.2> and <6.2.3>}

　　<6.4> $E_M^{current} \leftarrow E_M$;　{according to (5.8)}

　　<6.5> $epoch \leftarrow epoch + 1$;

　Until: $\left| E_M^{current} - E_M^{previous} \right| \leq \varepsilon$

End {MLP Algorithm – Training Phase}

After the training process of the MLP is finished, the variable *epoch* will contain the number of epochs required for completing the training of the network, in other words, how many times it was necessary to present all training samples to the network in order to adjust the weight matrices. The MLP will be considered completely trained (adjusted) when the mean squared error $\{E_M\}$ between two successive epochs is smaller than the precision $\{\varepsilon\}$, which is required for the problem being mapped.

The variable *epoch* can also be used as stop criterion for the training process of the MLP when the precision specified for a given problem becomes unreachable. For such purpose, the training process is simply stopped when the number of epochs reaches a predetermined number.

After the training process, the network can be used for estimating the outputs of the system when new samples are presented as inputs. The sequence of steps for this operation phase are as follows:

Begin {MLP Algorithm – Operation Phase}

<1> Obtain a sample {*x*};

<2> Assume $W_{ji}^{(1)}$, $W_{ji}^{(2)}$ and $W_{ji}^{(3)}$ already adjusted in the training stage;

<3> Execute the following instructions:

    <3.1> Obtain $I_j^{(1)}$ and $Y_j^{(1)}$; {according to (5.1) and (5.4)}

    <3.2> Obtain $I_j^{(2)}$ and $Y_j^{(2)}$; {according to (5.2) and (5.5)}   Forward steps

    <3.3> Obtain $I_j^{(3)}$ and $Y_j^{(3)}$; {according to (5.3) and (5.6)}

<4> Publish the outputs of the network, which are given by the elements of $Y_j^{(3)}$.

End {MLP Algorithm – Operation Phase}

It is important to note that the adjustment of the weight matrices is performed only in the training phase, in which the backward and forward stages are used to execute corrections to the synaptic weights. On the operation phase, no adjustment is performed in the internal parameters of the network, since only the forward stage is processed to generate the network outputs.

5.3.3 Optimized Versions of the Backpropagation Algorithm

Several variations of the backpropagation method have been proposed in order to enhance the efficiency of its convergence. Among these variations, one can find the method that uses the momentum parameter, resilient-propagation, and Levenberg-Marquardt methods.

(A) Method Using the Momentum Parameter

The momentum parameter is one of the simplest variations to implement in the backpropagation algorithm, since it only requires the addition of a single term in the adjustment equation, which will ponder upon how much the synaptic matrices changed between two successive iterations. Formally, considering the neurons belonging to the *L*th layer, we have:

$$W_{ji}^{(L)}(t+1) = W_{ji}^{(L)}(t) + \underbrace{\alpha \cdot \left(W_{ji}^{(L)}(t) - W_{ji}^{(L)}(t-1) \right)}_{\text{momentum term}} + \underbrace{\eta \cdot \delta_j^{(L)} \cdot Y_i^{(L-1)}}_{\text{learning term}}, \quad (5.41)$$

where α is defined as the momentum rate and its value is within the range of 0 and 1.

As it can be noticed from (5.41), when the momentum rate is equal to zero, the expression becomes equivalent to that of the conventional backpropagation. On the other hand, for values different from zero, the momentum term becomes relevant, and its contribution will positively affect the convergence process.

More specifically, when the current solution (reflected by its weight matrices) is far from the final solution (minimum point of the error function), the variation in the opposite direction of the gradient of the squared error function between two successive iterations will be significant. This implies that the difference between the error matrices of these two iterations will be relevant and, in this case, it is possible to perform a bigger incremental step for $W^{(L)}$ in the direction of the minimum of the error function. The momentum term is in charge of this task, since it is responsible for measuring this variation.

However, when the current solution is very near to the final solution, the variations on the weight matrices will be small, since the variation of the mean squared error between two successive iterations will be minor and, consequently, the contribution of the momentum term for the convergence process will be slight. From this moment on, all adjustments on the weight matrices are conducted (usually) only by the learning term.

Figure 5.8 illustrates the contribution of the Momentum Term (MT) and the Learning Term (LT) with respect to the convergence to the minimum W^{OT} of the squared error function.

Thus, when using the momentum parameter, the convergence process of the network becomes more efficient, since it takes into account how far the current

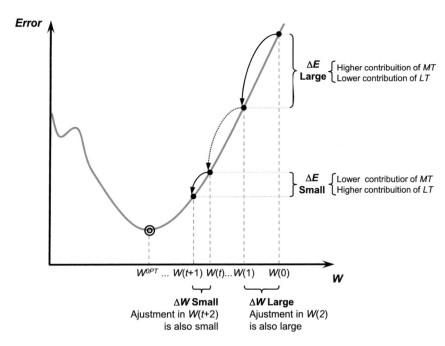

Fig. 5.8 Illustration of the training process using the momentum parameter method

solution is from the final solution (optimal). The use of the momentum parameter implies the acceleration of the convergence process with the rate of $\eta/(1 - \alpha)$ as analyzed in Reed and Mark II (1999). Values between the range of (0.05 $\leq \eta \leq$ 0.75) and (0 $\leq \alpha \leq$ 0.9) are usually recommended for the training of MLP networks (Rumelhart et al. 1986).

(B)   Resilient-Propagation Method

The logistic or hyperbolic tangent activation functions used in the neurons of the MLP produce limit values of 0 and 1 (for the logistic) and −1 and 1 (for the hyperbolic tangent) for the majority of points in their definition domain, as it can be observed in Fig. 5.9. Such circumstance implies the saturation of the outputs within these limit values if their activation potential $\{u\}$ have high values. Besides this, the partial derivatives $g'(u)$ would produce values next to zero, which in turn would slow down the training process, because, according to (5.11), (5.20), and (5.31), the backpropagation also depends on the calculation of $g'(u)$.

In this condition, the small variations in the gradient of the error function in (5.7) combined with the saturation intervals of the activation functions makes the convergence process slow. An additional computational effort will be required to guide the values of the weight matrices to the dynamic intervals of the given activation functions.

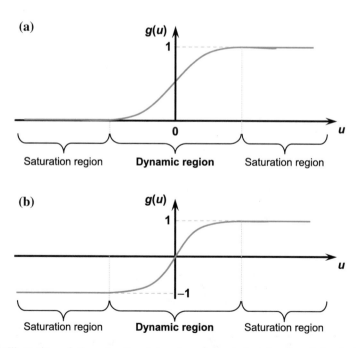

Fig. 5.9 Illustration of the saturation intervals and dynamic intervals of the logistic and hyperbolic tangent functions. **a** Activation function of logistic type. **b** Activation function of hyperbolic tangent type

Thus, the objective of the resilient-propagation method is to consider only the signal variations of the gradient of the error function (Riedmiller and Braun 1993), instead of considering the variations in its magnitude. Hence, the learning rate of this method becomes dynamic, because when the signals of the gradient are the same between two successive iterations, the learning rate can be incremented, since the convergence is distant from a minimum point (null gradient) of the error function. Otherwise, if the gradient signals are different, it implies the reduction of the learning rate in order to converge smoothly to the minimum point, while also taking into account the precision required for the problem.

An illustration of the convergence process made with the resilient-propagation method is presented in Fig. 5.10, in which the steps (I), (II), (III), and (V) imply positive increments (increasing) on the learning rate, since they have signal variations which are also positive; while steps (IV), (VI), and (VII) imply negative increments (decreasing), since their signal variations are also negative.

In mathematical notation, the evaluation of the change in the signal of the gradient is given by the following expressions:

$$\Lambda_{ji}^{(L)}(t) = \begin{cases} \eta^{+} \cdot \Lambda_{ji}^{(L)}(t-1), & \text{if } \frac{\partial E(t-1)}{\partial w_{ji}^{()}} \cdot \frac{\partial E(t)}{\partial w_{ji}^{()}} > 0 \\[2ex] \eta^{-} \cdot \Lambda_{ji}^{(L)}(t-1), & \text{if } \frac{\partial E(t-1)}{\partial w_{ji}^{()}} \cdot \frac{\partial E(t)}{\partial w_{ji}^{()}} < 0 \ , \\[2ex] \Lambda_{ji}^{(L)}(t-1), & \text{otherwise} \end{cases} \qquad (5.42)$$

where $\Lambda_{ji}^{(L)}$ is considered the individual learning rate associated with each element of $W_{ji}^{(L)}$, corresponding to the Lth weight matrix, with $(0 < \eta^{-} < 1)$ and $(\eta^{+} > 1)$

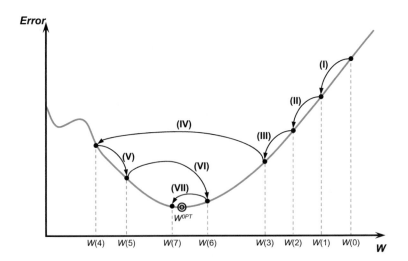

Fig. 5.10 Illustration of the convergence mechanism of the resilient-propagation method

being the constants responsible for incrementing or decrementing the individual learning rates of each weight matrix.

Finally, the weight matrices of the MLP are modified as follows:

$$
\Delta W_{ji}^{(L)}(t) =
\begin{cases}
-\Lambda_{ji}^{(L)}(t), & \text{if } \frac{\partial E(t)}{\partial w_{ji}^{(\cdot)}} > 0 \\
+\Lambda_{ji}^{(L)}(t), & \text{if } \frac{\partial E(t)}{\partial w_{ji}^{(\cdot)}} < 0 \\
0, & \text{otherwise}
\end{cases}
\tag{5.43}
$$

Therefore, it is possible to see in (5.43) that the weight matrices of the network are modified by the signal variations in the values of the partial derivatives, without considering the values of their magnitudes

(C) Levenberg-Marquardt Method

As described earlier, the backpropagation algorithm adjusts the values of the weight matrices in the opposite direction of the gradient of the squared error function. However, in practice, this algorithm tends to converge very slowly, thus requiring high computational effort. To avoid this inconvenience, several optimization techniques have been incorporated to the backpropagation algorithm to reduce its convergence time and mitigate the computational effort required. Among the most used optimization techniques for this purpose, the Levenberg-Marquardt algorithm (Hagan and Menhaj 1994) stands out.

The Levenberg-Marquardt algorithm is a second-order gradient method, based on the least squares method for nonlinear models, which can be incorporated into the backpropagation algorithm as to enhance the efficiency of the training process. For this algorithm, the squared error and mean squared error functions, provided respectively by (5.7) and (5.8), can be expressed together by:

$$
\begin{aligned}
V &= \frac{1}{2p} \cdot \sum_{k=1}^{p} \sum_{j=1}^{n_3} \left(d_j(k) - Y_j^{(3)}(k) \right)^2 \\
V &= \frac{1}{2p} \cdot \sum_{k=1}^{p} \left(d(k) - Y^{(3)}(k) \right)^T \cdot \left(d(k) - Y^{(3)}(k) \right) \\
V &= \frac{1}{2p} \cdot \sum_{k=1}^{p} E^T(k) \cdot E(k),
\end{aligned}
\tag{5.44}
$$

where the term $\{ E(k) = d(k) - Y^{(3)}(k) \}$ denotes the error vector with respect to the kth training sample. For a particular sample k, the error is obtained by:

$$
V = \frac{1}{2} \cdot E^T(k) \cdot E(k)
\tag{5.45}
$$

While the backpropagation algorithm is a method that decreases through the gradient of the squared error function to minimize it, the Levenberg-Marquardt algorithm is an approximation of the Newton method (Battiti 1992; Foresee and Hagan 1997). As such, the minimization of a function $V(z)$ with respect to a parametric vector z is given by the following iterative procedure:

$$\Delta z = -\left(\nabla^2 V(z)\right)^{-1} \cdot \nabla V(z), \tag{5.46}$$

where $\nabla^2 V(z)$ denotes the Hessian matrix (matrix of second-order derivatives) and $\nabla V(z)$ is the Jacobian matrix (matrix of first-order derivatives) of $V(z)$. Assuming that $V(z)$ is a function that executes m quadratic functions, such as those presented in (5.44), for a given parametric vector z composed of q elements, we have the following expression:

$$V(z) = \sum_{i=1}^{m} e_i^2(z) \tag{5.47}$$

Thus, from the previous equation, we can show that:

$$\nabla V(z) = J^T(z) \cdot e(z) \tag{5.48}$$

$$\nabla^2 V(z) = J^T(z) \cdot J(z) + \mu \cdot I, \tag{5.49}$$

where I is the identity matrix, μ is a parameter that adjusts the convergence rate of the Levenberg-Marquardt algorithm and $J(z)$ is the Jacobian matrix, which is defined by:

$$J(z) = \begin{bmatrix} \dfrac{\partial e_1(z)}{\partial z_1} & \dfrac{\partial e_1(z)}{\partial z_2} & \cdots & \dfrac{\partial e_1(z)}{\partial z_q} \\ \dfrac{\partial e_2(z)}{\partial z_1} & \dfrac{\partial e_2(z)}{\partial z_2} & \cdots & \dfrac{\partial e_2(z)}{\partial z_q} \\ \vdots & \vdots & \ddots & \vdots \\ \dfrac{\partial e_N(z)}{\partial z_1} & \dfrac{\partial e_N(z)}{\partial z_2} & \cdots & \dfrac{\partial e_N(z)}{\partial z_q} \end{bmatrix} \tag{5.50}$$

Inserting the results of (5.48) and (5.49) in (5.46), we obtain the iterative expression of the Levenberg-Marquardt method, that is:

$$\Delta z = \left(J^T(z) \cdot J(z) + \mu \cdot I\right)^{-1} \cdot J^T(z) \cdot e(z) \tag{5.51}$$

Therefore, the primary characteristic of this algorithm is the computation of the Jacobian matrix. For the training process of MLP networks, as illustrated in Fig. 5.3, this Jacobian matrix (5.50) is rewritten with respect to the synaptic matrices of the network, namely:

$$J(W) = \left[J\left(W^{(1)}\right) \quad J\left(W^{(2)}\right) \quad J\left(W^{(2)}\right) \right], \tag{5.52}$$

being $J(W) \in \Re^{(p)_x\left((n+1)\cdot n_1 + (n_1+1)\cdot n_2 + (n_2+1)\cdot n_3\right)}$. In this case, W is composed by:

$$
\begin{aligned}
W &= \left[W^{(1)} \quad W^{(2)} \quad W^{(3)} \right] \\
&= \begin{aligned}[t]
& [W_{1,0}^{(1)}\dots W_{1,n}^{(1)} \quad W_{2,0}^{(1)}\dots W_{2,n}^{(1)} \quad \cdots \quad W_{n_1,0}^{(1)}\dots W_{n_1,n}^{(1)} \\
& \ W_{1,0}^{(2)}\cdots W_{1,n_1}^{(2)} \quad W_{2,0}^{(2)}\cdots W_{2,n_1}^{(2)} \quad \cdots \quad W_{n_2,0}^{(2)}\cdots W_{n_2,n_1}^{(2)} \\
& \ W_{1,0}^{(3)}\cdots W_{1,n_2}^{(3)} \quad W_{2,0}^{(3)}\cdots W_{2,n_2}^{(3)} \quad \cdots \quad W_{n_3,0}^{(3)}\cdots W_{n_3,n_2}^{(3)}]^T,
\end{aligned}
\end{aligned}
\tag{5.53}
$$

where $W \in \Re^{\left((n+1)\cdot n_1 + (n_1+1)\cdot n_2 + (n_2+1)\cdot n_3\right)}$.

The matrices $J\left(W^{(1)}\right)$, $J\left(W^{(2)}\right)$ and $J\left(W^{(3)}\right)$ are, consequently, defined by:

$$
J\left(W^{(1)}\right) = \begin{bmatrix}
\frac{\partial E(1)}{\partial W_{1,1}^{(1)}} \cdots \frac{\partial E(1)}{\partial W_{1,n}^{(1)}} & \frac{\partial E(1)}{\partial W_{2,1}^{(1)}} \cdots \frac{\partial E(1)}{\partial W_{2,n}^{(1)}} & \cdots & \frac{\partial E(1)}{\partial W_{n_1,1}^{(1)}} \cdots \frac{\partial E(1)}{\partial W_{n_1,n}^{(1)}} \\
\frac{\partial E(2)}{\partial W_{1,1}^{(1)}} \cdots \frac{\partial E(2)}{\partial W_{1,n}^{(1)}} & \frac{\partial E(2)}{\partial W_{2,1}^{(1)}} \cdots \frac{\partial E(2)}{\partial W_{2,n}^{(1)}} & \cdots & \frac{\partial E(2)}{\partial W_{n_1,1}^{(1)}} \cdots \frac{\partial E(2)}{\partial W_{n_1,n}^{(1)}} \\
\vdots & \vdots & \ddots & \vdots \\
\frac{\partial E(p)}{\partial W_{1,1}^{(1)}} \cdots \frac{\partial E(p)}{\partial W_{1,n}^{(1)}} & \frac{\partial E(p)}{\partial W_{2,1}^{(1)}} \cdots \frac{\partial E(p)}{\partial W_{2,n}^{(1)}} & \cdots & \frac{\partial E(p)}{\partial W_{n_1,1}^{(1)}} \cdots \frac{\partial E(p)}{\partial W_{n_1,n}^{(1)}}
\end{bmatrix}
\tag{5.54}
$$

$$
J\left(W^{(2)}\right) = \begin{bmatrix}
\frac{\partial E(1)}{\partial W_{1,1}^{(2)}} \cdots \frac{\partial E(1)}{\partial W_{1,n_1}^{(2)}} & \frac{\partial E(1)}{\partial W_{2,1}^{(2)}} \cdots \frac{\partial E(1)}{\partial W_{2,n_1}^{(2)}} & \cdots & \frac{\partial E(1)}{\partial W_{n_2,1}^{(2)}} \cdots \frac{\partial E(1)}{\partial W_{n_2,n_1}^{(2)}} \\
\frac{\partial E(2)}{\partial W_{1,1}^{(2)}} \cdots \frac{\partial E(2)}{\partial W_{1,n_1}^{(2)}} & \frac{\partial E(2)}{\partial W_{2,1}^{(2)}} \cdots \frac{\partial E(2)}{\partial W_{2,n_1}^{(2)}} & \cdots & \frac{\partial E(2)}{\partial W_{n_2,1}^{(2)}} \cdots \frac{\partial E(2)}{\partial W_{n_2,n_1}^{(2)}} \\
\vdots & \vdots & \ddots & \vdots \\
\frac{\partial E(p)}{\partial W_{1,1}^{(2)}} \cdots \frac{\partial E(p)}{\partial W_{1,n_1}^{(2)}} & \frac{\partial E(p)}{\partial W_{2,1}^{(2)}} \cdots \frac{\partial E(p)}{\partial W_{2,n_1}^{(2)}} & \cdots & \frac{\partial E(p)}{\partial W_{n_2,1}^{(2)}} \cdots \frac{\partial E(p)}{\partial W_{n_2,n_1}^{(2)}}
\end{bmatrix}
\tag{5.55}
$$

$$
J\left(W^{(3)}\right) = \begin{bmatrix}
\frac{\partial E(1)}{\partial W_{1,1}^{(3)}} \cdots \frac{\partial E(1)}{\partial W_{1,n_2}^{(3)}} & \frac{\partial E(1)}{\partial W_{2,1}^{(3)}} \cdots \frac{\partial E(1)}{\partial W_{2,n_2}^{(3)}} & \cdots & \frac{\partial E(1)}{\partial W_{n_3,1}^{(3)}} \cdots \frac{\partial E(1)}{\partial W_{n_3,n_2}^{(3)}} \\
\frac{\partial E(2)}{\partial W_{1,1}^{(3)}} \cdots \frac{\partial E(2)}{\partial W_{1,n_2}^{(3)}} & \frac{\partial E(2)}{\partial W_{2,1}^{(3)}} \cdots \frac{\partial E(2)}{\partial W_{2,n_2}^{(3)}} & \cdots & \frac{\partial E(2)}{\partial W_{n_3,1}^{(3)}} \cdots \frac{\partial E(2)}{\partial W_{n_3,n_2}^{(3)}} \\
\vdots & \vdots & \ddots & \vdots \\
\frac{\partial E(p)}{\partial W_{1,1}^{(3)}} \cdots \frac{\partial E(p)}{\partial W_{1,n_2}^{(3)}} & \frac{\partial E(p)}{\partial W_{2,1}^{(3)}} \cdots \frac{\partial E(p)}{\partial W_{2,n_2}^{(3)}} & \cdots & \frac{\partial E(p)}{\partial W_{n_3,1}^{(3)}} \cdots \frac{\partial E(p)}{\partial W_{n_3,n_2}^{(3)}}
\end{bmatrix},
\tag{5.56}
$$

where $J\left(W^{(1)}\right) \in \Re^{(p)\times(n\cdot n_1)}$, $J\left(W^{(2)}\right) \in \Re^{(p)\times(n_1\cdot n_2)}$ and $J\left(W^{(3)}\right) \in \Re^{(p)\times(n_2\cdot n_3)}$.

From (5.51), the iterative expression of the Levenberg-Marquardt method that adjusts the MLP weight matrices is redefined by:

$$\Delta W = \left[J^T(W) \cdot J(W) + \mu I\right]^{-1} \cdot J^T(W) \cdot E, \qquad (5.57)$$

where $E = [E(1)\, E(2) \cdots E(p)]^T$ is the error vector with respect to the p training samples.

Finally, the elements of the matrices $J\left(W^{(1)}\right)$, $J\left(W^{(2)}\right)$ and $J\left(W^{(3)}\right)$ are obtained, in sequence, from the forward and backward stages employed in the conventional backpropagation algorithm, which was presented in the previous subsection. By implementing these modifications, it is proved that Levenberg-Marquardt method can outperform the MLP training process in the order of 10–100 times faster than the conventional backpropagation algorithm (Hagan and Menhaj 1994). However, problems of convergence can occur if matrix J (z) used in (5.51) is ill-conditioned.

5.4 Multilayer Perceptron Applications

Multilayer Perceptron networks can be considered the most used network to solve problems from knowledge areas related to sciences and engineering. In fact, it is possible to find applications of MLP networks in several areas of knowledge, such as medicine, biology, chemistry, physics, economy, geology, ecology, and psychology, and numerous applications in the most different themes involving engineering.

Considering the range of applications in which MLP networks can be employed, there are three classes of problem that gather most of its applications, which are: problems involving pattern classification (recognition); problems related to function approximation; and problems related to dynamic (time-variant) systems. Due to its relevance, these three classes of problems will be addressed separately in the following subsection.

5.4.1 Problems of Pattern Classification

As we highlighted in Chap. 1, a problem of pattern classification consists in associating an input pattern (sample) to one of the previously defined classes of the problem. As an example, it is possible to create an application where the MLP is trained for recognizing voice signals in order to allow people into restrict-access environments. In this scenario, considering that the training was performed with vocables of only three persons, the network response for a voice signal inserted in its inputs would inform who the owner of the given signal is.

Another relevant aspect that can be observed from this simple example is that the outputs associated with problems of pattern classification are always discrete magnitudes (enumerable). The most elementary scenarios would be those of binary

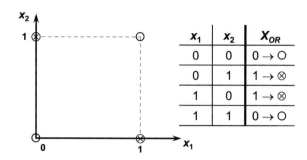

Fig. 5.11 Illustration of the exclusive-or problem

x_1	x_2	x_{OR}
0	0	$0 \to \bigcirc$
0	1	$1 \to \otimes$
1	0	$1 \to \otimes$
1	1	$0 \to \bigcirc$

outputs, in which there are only two classes of possible responses, and both could be representing, for example, the "presence" or "absence" of a given attribute in a sample presented to the network. Thus, considering that the network output provides numerical responses, one possible codification would be to assign the value 0 to the attribute "presence" while the value 1 would be assigned to the attribute "absence." As will be discussed later in the book, a system similar to this could also be used for multi-class (with three or more classes) problems.

Regarding the problems with binary outputs, according to Chap. 2, the Perceptron (single layer) could only converge if the two classes associated with the problem were linearly separable. Otherwise, the Perceptron would never converge to place its hyperplane within the range delimited by the separability boundary of the classes. A classic case of such fact is found in the exclusive-or (*Xor* gate) problem of the Boolean algebra, as illustrated in Fig. 5.11.

By analyzing the graphical representation in Fig. 5.11, it is possible to observe that it would be impossible to place a single straight line that could separate both classes of the exclusive-or problem.

Other similar scenarios can only be solved by an MLP network with two neural layers, such as that presented in Fig. 5.12.

To understand the mechanisms involved with that pattern recognition problem, Fig. 5.13 illustrates a configuration of (separability) lines that will be implemented by neurons *A* and *B*, from the topological configuration in Fig. 5.12, using the backpropagation algorithm.

In such example, assuming the logistic function in each neuron of the MLP of Fig. 5.12, it is possible to see that the neuron *A* will have its output equal to 1 only for those patterns that are above its boundary line, while the neuron *B* will provide the value 1 for all patterns that are below its boundary line. As a consequence, neuron *Y* will produce as an output value equal to 1 only for the cases in which the

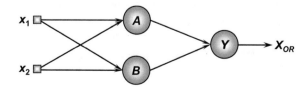

Fig. 5.12 An MLP network applied to the exclusive-or problem

Fig. 5.13 The exclusive-or
problem

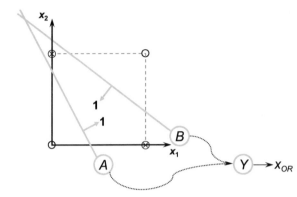

Fig. 5.14 Decision boundary
of the exclusive-or problem

Response of the neuron "A"

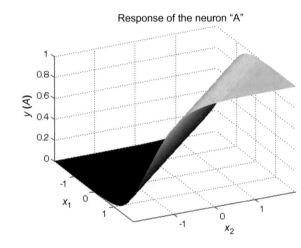

responses from neurons A and B are also equal to 1; otherwise, its output value will
be equal to 0. Therefore, it is important to note that the neuron Y is simply performing a Boolean operation of conjunction (AND gate).

Figure 5.14 illustrates the decision surface generated by the logistic function of
neuron A, in which it is possible to see that the value 1 is produced only when
$(x_1 = 0)$ and $(x_2 = 1)$.

Figure 5.15 illustrates the combination of the logistic functions made by the
neuron Y. It is possible to verify that neuron Y have inverted the logistic function
related to neuron B in order to correctly classify the classes of the exclusive-or
problem.

Finally, Fig. 5.16 shows the boundary lines associated with the exclusive-or
problem. They are delimited by two straight lines obtained by the intersection of the
surface illustrated in Fig. 5.15 with the plane $y = 0$.

In a similar fashion, it is possible to deduce that an MLP with two neural layers,
where one layer is the hidden layer, and the other is the output layer, can map any

Fig. 5.15 Illustration of the output of neuron Y combining both logistic functions from neurons A and B

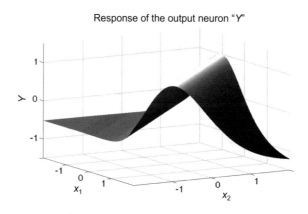

Response of the output neuron "Y"

Fig. 5.16 Classification borders for the exclusive-or problem

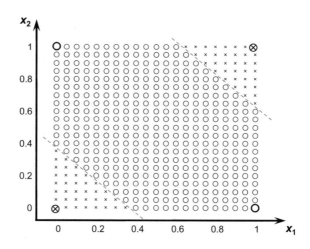

pattern classification problem whose elements are within a convex region, such as those illustrate in Fig. 5.17.

In such examples, the classification of the samples in Fig. 5.17a would require two neurons in the hidden layer of the MLP, whereas the classification of the samples in Fig. 5.17b, c would require six and four neurons on the hidden layers of the MLP networks

From a geometric point of view, a region is considered convex if, and only if, all the points of any line segment defined between any pair of points from the domain are inside this region. Figure 5.18a presents an illustration of a convex region while Fig. 5.18b shows a non-convex region.

Thus, considering that MLP networks with a single hidden layer can classify patterns placed within a convex region, it is possible to deduce that MLP networks with two hidden layers can classify patterns that are within any geometric region (Lui 1990; Lippmann 1987), including non-convex regions such that showed in Fig. 5.19.

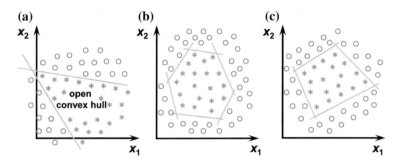

Fig. 5.17 Examples of convex regions of pattern classification problems

Fig. 5.18 Illustration of a
convex region and a
non-convex region

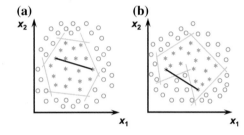

Fig. 5.19 Decision
boundaries of a problem
within a non-convex region

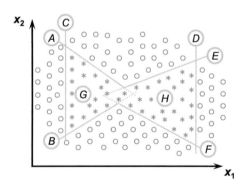

For such case, the configuration of the MLP network showed in Fig. 5.20 represents a topology that can implement the pattern classification task required by the problem depicted in Fig. 5.19.

In this case, it is possible to consider, for instance, that neurons *A*, *B* and *C* of the first hidden layer are responsible for delimiting the left convex region (triangle), while the neurons *D*, *E* and *F* are related to the right convex region (triangle). Neuron *G* of the second middle layer is responsible for combining the outputs of neurons *A*, *B*, and *C* in order to represent the group that belongs to the left convex region so that in this condition its output would be equal to 1. In a similar manner,

Fig. 5.20 Example of an MLP network applied to pattern classification problems within non-convex regions

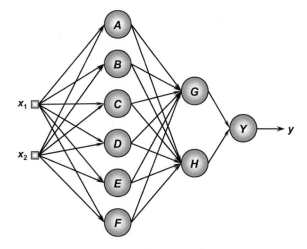

neuron H would produce a value equal to 1 in order to represent the right convex region, whose geometry surrounds the patterns belonging to the that same class.

Finally, neuron Y of Fig. 5.20 is responsible for performing the Boolean operation of disjunction (OR gate), because if one of the outputs produced by neurons G or H is equal to 1, its final response y should also be equal to 1.

Additionally, the MLP with two hidden layers could also map other types of spatial regions, such as those formed by disjunctive sets (disjointed), as illustrated in Fig. 5.21.

On the other hand, although an MLP constituted by two hidden layers is sufficient for recognizing patterns that are delimited by any geometric region, there are scenarios where networks with more than two hidden layers are employed. Such configurations can be more appropriated for both enhancing the performance of the training process and reducing the topology of the network.

Besides that, there are other particular scenarios, such as those reported in Makhoul et al. (1989), Liu (1990), in which MLP networks with a single hidden

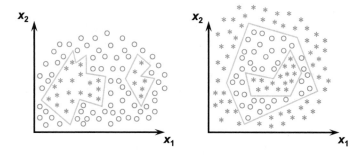

Fig. 5.21 Examples of disjoint regions in pattern classification

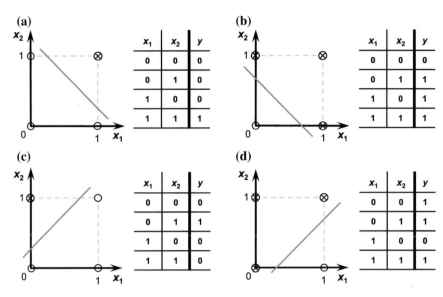

Fig. 5.22 Examples of linearly separable logical operations. **a** $y \leftarrow (x_1)\text{AND}(x_2)$. **b** $y \leftarrow (x_1)\text{OR}(x_2)$. **c** $y \leftarrow (\text{NOT}x_1)\text{AND}(x_2)$. **d** $y \leftarrow (\text{NOT}x_1)\text{OR}(x_2)$

layer are still capable of mapping problems whose patterns are placed within disjoint or non-convex regions.

In summary, it is possible to conclude that when an MLP is applied to pattern classification problems, the output neurons of the network perform logical combinations, such as the AND and OR operations, to the regions that were defined by the neurons in previous layers, independently from the dimension of the input patterns. These logical operations can be used and mapped because they are linearly separable. Figure 5.22 illustrates some of these (linearly separable) logical operations which can be implemented by the output neurons. Thus, considering two logical variables x_1 and x_2, there are 16 Boolean operations possible to be made, among which only two of them (the exclusive-or and its complement) are not linearly separable.

Also, in the case of pattern classification problems with more than two classes, there is the need for inserting more neurons on the output layer of the network, because an MLP with a single neuron in its output layer can distinguish just two classes. As an example, an MLP composed of two neurons in its output layer could represent, at most, four classes (Fig. 5.23). A network with three neurons in the output layer could classify a total of eight classes. Generalizing this concept, an MLP with m neurons in its output layer would be able to classify, theoretically, up to 2^m classes.

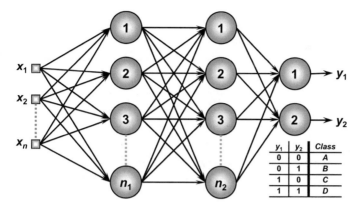

Fig. 5.23 Sequential binary codification for pattern classification problems involving four possible classes

However, depending on the complexity of the problem being analyzed, by adopting this binary codification, the MLP training processes can become harder (Hampshire II and Pearlmutter 1991), since the classes would be represented by points that are spatially next to each other. Such scenario could demand a substantial increase in the number of neurons in the intermediate layers, as well as an eventual difficulty in adjusting the topology of the network.

Alternatively, one of the most used methods of codification is the "one of c-class", which consists in associating the output of each neuron directly to the class. Hence, the number of neurons in the output layer will be the same as the number of classes of the problem. In a problem with four classes, the output configuration of an MLP network would also be composed of four neurons, as illustrated in Fig. 5.24.

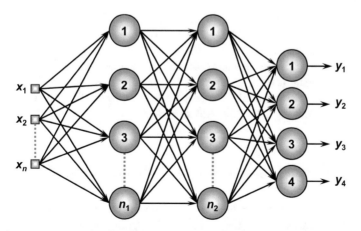

Fig. 5.24 One of c-classes codification method for a pattern classification problem with four possible classes

Table 5.1 Codification of classes for pattern classification problems	Binary sequence codification			"one of c-classes" codification				
	y_1	y_2	Class	y_1	y_2	y_3	y_4	Class
	0	0	A	1	0	0	0	A
	0	1	B	0	1	0	0	B
	1	0	C	0	0	1	0	C
	1	1	D	0	0	0	1	D

As an example, Table 5.1 describes the codification of these four classes, considering both the binary sequence codification (Fig. 5.23) and "one of c-classes" method (Fig. 5.24).

About the "one of c-classes", by inspecting Table 5.1, it is possible to verify that a given sample would belong to class B only if the output of neuron y_2 was equal 1, while all other outputs were equal to 0. As stated before, such strategy of orthogonal codification became known as the "one of c-classes" method (Duda et al. 2001) in which it is assumed that each class of the problem will be represented by a single neuron.

Finally, when a network is used for solving pattern classification problems, the output values y_i provided by the neurons of the output layer must be post-processed, since the activation functions produce real numbers that, in the case of the logistic function, might be close to the zero or one. In this case, depending on the required precision for the solutions, the values of y_i^{post} arisen by this post-processing operation can be obtained with the following expression:

$$y_i^{post} = \begin{cases} 1, & \text{if } y_i \geq \text{threshold}^{sup} \\ 0, & \text{if } y_i \leq \text{threshold}^{inf} \end{cases}, \tag{5.58}$$

where threshold^{sup} and threshold^{inf} define, respectively, the superior and inferior threshold values for the neurons of the output layer. That is, if the output of a neuron is greater than or equal to threshold^{sup}, then the value 1 is assumed; otherwise, if its smaller or equal to threshold^{inf}, it will assume the value 0. The specification of such limits depends essentially on the required precision. In the case of using the logistic activation function, the usual values adopted for threshold^{sup} and threshold^{inf} are $\text{threshold}^{sup} \in [0.5; 0.9]$ and $\text{threshold}^{inf} \in [0.1; 0.5]$. Other implementation details will be presented in Sect. 5.9.

5.4.2   Functional Approximation Problems (Curve Fitting)

Another type of problem in which the MLP network is majorly used is the functional approximation problem. It consists of mapping the behavior of a process based on several measurements of its inputs and outputs. In fact, it is possible to observe through these problems one of the intrinsic features of artificial neural networks, which is, learning by examples. In the case of functional approximation, this feature is translated as having available a set of inputs/outputs that reproduce the behavior of the system being mapped.

Due to this ability of mapping processes through examples, MLP networks become candidates for the solution of several applications where the only information available is a set of inputs/outputs. Artificial neural networks have been extensively employed in scenarios where the process being modeled is complex, and the application of conventional methods provide unsatisfactory results; or in scenarios where the already modeled system becomes overly particularized around some operation points that produce satisfactory results.

The universal approximation theorem for MLP networks, (Kolmogorov 1957), provides the necessary basis for defining the structural configurations of these networks for mapping algebraic functions (Cybenko 1989).

Assuming that the activation function $g(.)$ of the MLP networks are continuous functions and limited on their images, such as the logistic and hyperbolic tangent function, it is possible to demonstrate that an MLP topology composed of only a single hidden neural layer is able to map any function that is continuous in the subspace of the real functions. In mathematical notation, we have:

$$y(x_1, x_2, \ldots, x_n) = \sum_{i=1}^{n_1} \underbrace{\lambda_i}_{\text{parcel (i)}} \cdot \underbrace{g_i^{(1)}\left(u_i^{(1)}\right)}_{\text{parcel (ii)}} \qquad (5.59)$$

$$u_i^{(1)} = \sum_{j=1}^{n} W_{ji}^{(1)} \cdot x_j - \theta_i \, , \qquad (5.60)$$

where λ_i are constants that weigh functions $g_i^{(1)}(.)$

The expressions (5.59) and (5.60) can be translated into an MLP representation, such as that in Fig. 5.25, which is composed, in fact, of a single hidden neural layer whose neurons use the logistic function (1.8) as its activation function. In other words, the function y to be fitted by the MLP will be composed of a superposition of logistic activation functions {parcel (ii)}, represented by the terms $g_i^{(1)}\left(u_i^{(1)}\right)$, which are weighted by the factors λ_i {parcel (i)}. Similarly, the demonstration is equally valid when the hyperbolic tangent is assumed as the activation functions of the neurons of the hidden layer.

In sequence, using the linear function (1.11) as the activation function of the output layer, the single output neuron will perform a linear combination of the logistic activation functions of the neurons in the previous layer.

Therefore, after the training process of the MLP network is finished, the weight matrix corresponding to the output neuron will relate to the values λ_i of expression (5.59), that is, $\lambda_i = W_{1,i}^{(2)}$.

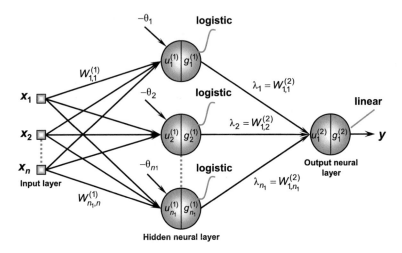

Fig. 5.25 Illustration of an MLP network for function approximation

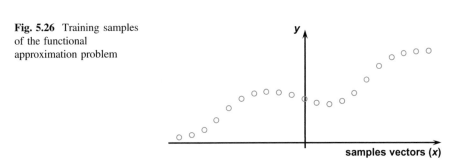

Fig. 5.26 Training samples of the functional approximation problem

As an example, consider an MLP to implement the fitting of the curve in Fig. 5.26.

In this case, an MLP will map (approximate) the functional behavior of the process. As an example of topology, an MLP with three neurons {A, B, C} in its hidden layer will be used, as illustrated in Fig. 5.27.

After adjusting the internal parameters of the MLP through the backpropagation algorithm, Fig. 5.28 presents a possible solution obtained from the linear combination of the (nonlinear) logistic activation functions that compose the output of the neurons in the hidden layer. Indeed, it is possible to observe that the output neuron implements an operation that can map the behavior illustrated in Fig. 5.26.

From Fig. 5.28, it is also possible to observe that the thresholds {θ} of neurons A, B, and C, of to the hidden layer, are responsible for translating the activation functions within its definition domains, while the weights {λ_i} of the output neuron Y are responsible for scaling these activation functions.

Similarly to pattern classification problems, although an MLP with a single hidden layer is sufficient for mapping any nonlinear continuous function defined in

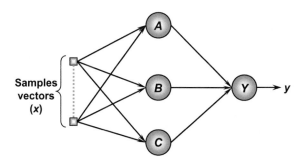

Fig. 5.27 MLP topology used in the problem of functional approximation

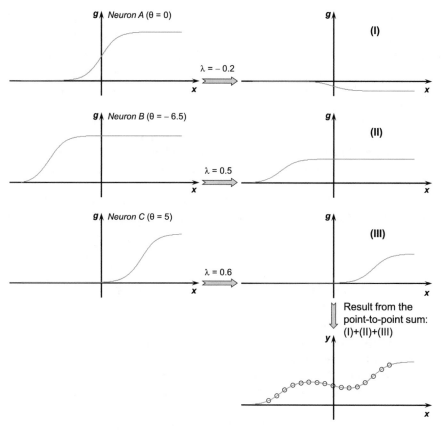

Fig. 5.28 Illustration of the superposition of logistic activation functions for a curve fitting problem

a compact (closed) domain, there are situations where more than one single hidden layer is used to increase the performance of the training process and to reduce its topological structure.

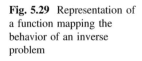

Fig. 5.29 Representation of a function mapping the behavior of an inverse problem

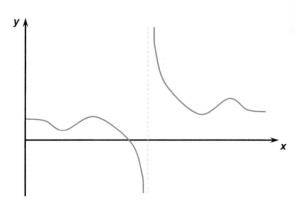

Moreover, it is important to stress that the universal approximation theorem states that only one hidden layer is needed; however, the required number of neurons in this layer to perform such task is yet unknown and, depending on the complexity of the problem being mapped, an expressive number of neurons may be required.

Additionally, for some particular problems within the class of inverse problems, such as those involving robot inverse kinematics, it is proved that its mapping is only possible if made through MLP topologies with two hidden layers (Sontag 1992). Figure 5.29 illustrates a function obtained from the mapping of an inverse problem.

As observed in Fig. 5.29, it is possible to see a discontinuity in the function domain. This condition requires the use of two hidden layers in the network when the MLP is applied to inverse problems.

5.4.3 Problems Involving Time-Variant Systems

The last category of problems to be addressed in this book, concerning the use of MLP networks, is known as dynamic systems, whose behaviors are considered time-variant or time-dependent.

As examples, consider a weekly forecast of future values of stocks in the financial market, or the prediction of electrical energy consumption for the next months.

Differently from functional approximation and pattern recognition problems (considered static), the outputs of dynamic systems, at any instant of time, depend on their previous input and output values (Ljung 1999).

To stress the differences between functional approximation problems and those related to dynamic systems, Fig. 5.30 illustrates the domain of definition of the training/test data used by an MLP in a functional approximation problem, which are delimited by the minimum value $\{x_i^{min}\}$ and maximum value $\{x_i^{max}\}$ of each input

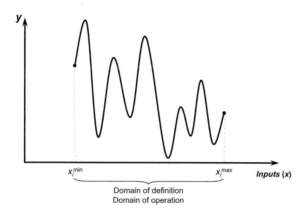

Fig. 5.30 Definition and operation domain of the MLP applied to problems of functional approximation

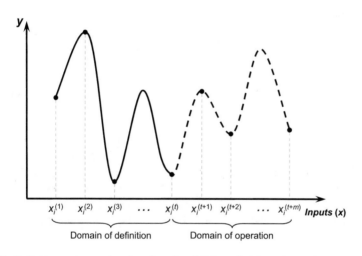

Fig. 5.31 Definition and operation domains of an MLP applied in problems involving dynamic systems

variable. In this case, after the training is complete, the domain of operation of the MLP network coincides with the domain of definition of the data, since the responses produced will always depend on the input values ranging from x_i^{min} to x_i^{max}.

On the other hand, Fig. 5.31 shows the definition and operation domains associated with an MLP that is mapping a dynamic system. It is possible to observe that both domains are governed by time, and the operation domain begins after the definition domain. In dynamic systems, the current output depends on the previous inputs and outputs, then the training/test data are used to adjust the internal parameters of the network. Thus, the network will be able to estimate future values that will belong to its operation domain.

Hence, when dealing with MLP for mapping problems that involve dynamic systems, two main configurations can be used, that is, the time delay neural network (TDNN) and the MLP with recurrent inputs.

Moreover, it must be considered that unlike MLP networks with two or more hidden layers, MLP networks with a single hidden layer are usually less prone to converge to a local minima, since their compact structures reduce the geometric complexity of the function that represents the mean squared error (Curry and Morgan 2006; Xiang et al. 2005).

(A) Time Delay Neural Network

Time delay neural networks were first idealized by Lang and Hinton (1988), and belong to the feedforward architectures with multiple layers, without any feedback from the neurons of posterior layers to the neurons of the first layer.

The forecast or prediction of values after a given time t is computed regarding previous values, that is:

$$x(t) = f\big(x(t-1), x(t-2), \ldots, x(t-n_p)\big), \tag{5.61}$$

where n_p is the order of the predictor, that is, the amount of past samples that will be required for the estimation of the value $x(t)$. In the terminology used in the system identification area, the model presented in (5.61) is also known as autoregressive (AR), whose function $f(.)$ will be implemented by the MLP network.

Thus, considering the expression (5.61), an MLP network applied to time-variant processes would have a topological configuration similar to that presented in Fig. 5.32. Differently from the original TDNN conception (Waibel et al. 1989), in which the temporal delays are inserted in all the layers of the network, the configuration illustrated in Fig. 5.32 introduces delays only in the input layer. Therefore, such topological arrangement is also known as Focused Time-Lagged

Fig. 5.32 Focused time-lagged feedforward network topology

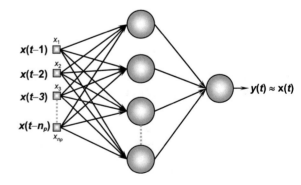

Feedforward Network (Haykin 2009). In this case, the time delays serve as a memory, guaranteeing that previous samples, which reflect the temporal behavior of the process, are always inserted in the network without the need to feedback its outputs.

From Fig. 5.32, it is possible to verify that the network receives the n_p inputs $\{x(t-1), x(t-2), \ldots, x(t-n_p)\}$, which represent the process behavior, and predicts the corresponding response expected for $x(t)$ through the output value y (t) provided by the output neuron. Thus, during the training process, the network will try to adjust its weight matrices in order to minimize the error $E(t)$ that is produced by the difference between $x(t)$ and $y(t)$. In mathematical terms, we have:

$$E(t) = x(t) - y(t), \quad \text{with } (n_p + 1) \leq t \leq N, \tag{5.62}$$

where n is the total number of available measurements (samples) that were sequentially collected over time.

The training of the Time-Lagged Feedforward Network is similar to that of the conventional MLP, and the learning process is identical. Some care must be taken when creating the training set. To demonstrate this mechanism, assume that eight measurements $\{N = 8\}$ were captured for a given dynamic system:

$$
\begin{array}{cccccccc}
& t=1 & t=2 & t=3 & t=4 & t=5 & t=6 & t=7 & t=8 \\
x(t) = & [0.11 & 0.32 & 0.53 & 0.17 & 0.98 & 0.67 & 0.83 & 0.79]^T \\
& x(1) & x(2) & x(3) & x(4) & x(5) & x(6) & x(7) & x(8)
\end{array} \tag{5.63}
$$

Assuming that the process will be mapped with a third-order prediction $\{n_p = 3\}$, the Time-Lagged Feedforward Network will have five patterns in its training set because, according to (5.62), the parameter t will vary from 4 to 8, as illustrated by the next table.

list of input/output					training set			
	x_1	x_2	x_3	desired output		x_1	x_2	x_3	d
$t=4$	$x(3)$	$x(2)$	$x(1)$	$x(4)$	$x^{(1)}$	0.53	0.32	0.11	$d^{(1)}=0.17$
$t=5$	$x(4)$	$x(3)$	$x(2)$	$x(5)$	$x^{(2)}$	0.17	0.53	0.32	$d^{(2)}=0.98$
$t=6$	$x(5)$	$x(4)$	$x(3)$	$x(6)$	$x^{(3)}$	0.98	0.17	0.53	$d^{(3)}=0.67$
$t=7$	$x(6)$	$x(5)$	$x(4)$	$x(7)$	$x^{(4)}$	0.67	0.98	0.17	$d^{(4)}=0.83$
$t=8$	$x(7)$	$x(6)$	$x(5)$	$x(8)$	$x^{(5)}$	0.83	0.67	0.98	$d^{(5)}=0.79$

With center column: $4 \leq t \leq 8$ (order 3) $n_p = 3$

where the value $x_0 = -1$, associated with the neuron threshold, must be considered in all of them.

A sliding window operation with width n_p is then performed on the vector x (t) (5.63), moving itself by one unit at each iteration. The following picture illustrates this mechanism.

$$x(t) = [\; \underbrace{0.11 \quad 0.32 \quad 0.53 \quad 0.17}_{\text{window 1 }(t=4)} \; 0.98 \quad 0.67 \quad 0.83 \quad 0.79 \;]^T$$

$$x(t) = [\; 0.11 \; \underbrace{0.32 \quad 0.53 \quad 0.17 \quad 0.98}_{\text{window 2 }(t=5)} \; 0.67 \quad 0.83 \quad 0.79 \;]^T$$

$$x(t) = [\; 0.11 \quad 0.32 \; \underbrace{0.53 \quad 0.17 \quad 0.98 \quad 0.67}_{\text{window 3 }(t=6)} \; 0.83 \quad 0.79 \;]^T$$

$$x(t) = [\; 0.11 \quad 0.32 \quad 0.53 \; \underbrace{0.17 \quad 0.98 \quad 0.67 \quad 0.83}_{\text{window 4 }(t=7)} \; 0.79 \;]^T$$

$$x(t) = [\; 0.11 \quad 0.32 \quad 0.53 \quad 0.17 \; \underbrace{0.98 \quad 0.67 \quad 0.83 \quad 0.79}_{\text{window 5 }(t=8)} \;]^T$$

After the training is completed, the network can be used to estimate future (posterior) values by just using previous samples. The network now has the operation domain $\{t \geq 9\}$. Consider the following example illustrated in the table:

prediction of future values

	x_1	x_2	x_3	estimated output
$t = 9$	$x(8)$	$x(7)$	$x(6)$	$x(9) \approx y(9)$
$t = 10$	$x(9)$	$x(8)$	$x(7)$	$x(10) \approx y(10)$
$t = 11$	$x(10)$	$x(9)$	$x(8)$	$x(11) \approx y(11)$
(...)	(...)	(...)	(...)	(...)

From the table above, it is possible to see that to predict the future behavior of the process at the first instant in the operation domain $\{t = 9\}$, it is required the input of the three last values from the series $\{x(8), x(7), x(6)\}$ to provide the estimative of $x(9)$, represented by the value $y(9)$ of the output neuron of the network. Consequently, to estimate the value of $x(10)$, the last two original values $\{x(8), x(7)\}$ and the estimated value of $x(9)$, produced by the last iteration, must be used.

Thus, it can be stated that the network always performs a prediction, computing its current or future values sequentially from the last three values, if a prediction of third-order $\{n_p = 3\}$ is assumed. However, there are situations where the prediction order must be increased in order to ensure better accuracy on the estimation of the future behavior of the process. For example, in the case of a prediction of fourth order $\{n_p = 4\}$, the next value obtained by the network would depend on the last four values, and its topology in this case would be composed of four inputs.

(B) **MLP with Feedback**

Differently from the MLP with time-delayed inputs, the architecture with recurrent outputs allows the recovering of previous responses from the feedback of signals produced by early moments. It can be said that such topologies have memory, thus being able to "remember" previous outputs so to produce the current or future response.

According to the definition of neural architectures from Sect. 2.2, these networks belong to the recurrent or feedback architectures. The prediction of future values associated with the behavior of the process, from a given time t, is based on the previous inputs and past values produced by its outputs, that is:

$$x(t) = f\big(x(t-1), x(t-2), \ldots, x(t-n_p), y(t-1), y(t-2), \ldots, y(t-n_q)\big),$$

$$(5.64)$$

where n_p is the predictor order and indicates the number of previous measurements (samples) that are necessary for the estimation of $x(t)$. The value n_q expresses the context order, that is, the quantity of past outputs that will be used on the estimation of $x(t)$. In this condition, the role performed by the network, after its training is complete, will be of representing indirectly the function $f(.)$, which is responsible for identifying the relationship among inputs and outputs of the system. As a consequence, Fig. 5.33 illustrates a representation of a recurrent MLP that implements the dynamic process of (5.64).

It is possible to verify, by analyzing Fig. 5.33, that the context signals are feeding back all n_q past outputs, produced at previous instants, to the neurons of the first layer.

Such configuration allows recurrent MLP networks to map the inputs and outputs of a process that may be nonlinear as well as time-variant, thus becoming a flexible tool for applications involving system identification.

The recurrent MLP that implements the relationship presented in (5.64) is a Nonlinear AutoRegressive eXogenous model (NARX), whose applicability is to map systems with typically nonlinear dynamics (Nelles 2010).

Hence, such as the MLP with time delay inputs, the training of recurrent MLP networks is performed in a similar way to that of the conventional MLP, which was studied for both pattern classification and functional approximation problems. Therefore, based on the expression (5.64), the training of the recurrent MLP promotes the required adjustments in its weight matrices in order to minimize the error $E(t)$ between the expected value of $x(t)$ given the response $y(t)$ estimated by the network.

Thereby, consider the temporal sequence provided by (5.63) and the topology presented in Fig. 5.33, the training set for the recurrent MLP when $n_p = 3$

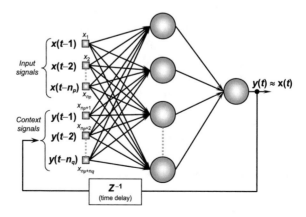

Fig. 5.33 MLP topology with the outputs being feedback to the inputs

(prediction of third-order, using three past inputs) and $n_q = 2$ (context of second-order, using two past outputs) would be constituted by:

list of input/output									training set					
	x_1	x_2	x_3	x_4	x_5	desired output	MLP output			x_1	x_2	x_3	x_4	x_5	d
$t=4$	$x(3)$	$x(2)$	$x(1)$	0	0	$x(4)$	$y(4)$	$4 \le t \le 8$	$x^{(1)}$	0.53	0.32	0.11	0	0	$d^{(1)} = 0.17$
$t=5$	$x(4)$	$x(3)$	$x(2)$	$y(4)$	0	$x(5)$	$y(5)$		$x^{(2)}$	0.17	0.53	0.32	$y(4)$	0	$d^{(2)} = 0.98$
$t=6$	$x(5)$	$x(4)$	$x(3)$	$y(5)$	$y(4)$	$x(6)$	$y(6)$	(order 3)	$x^{(3)}$	0.98	0.17	0.53	$y(5)$	$y(4)$	$d^{(3)} = 0.67$
$t=7$	$x(6)$	$x(5)$	$x(4)$	$y(6)$	$y(5)$	$x(7)$	$y(7)$	$n_p = 3$	$x^{(4)}$	0.67	0.98	0.17	$y(6)$	$y(5)$	$d^{(4)} = 0.83$
$t=8$	$x(7)$	$x(6)$	$x(5)$	$y(7)$	$y(6)$	$x(8)$	$y(8)$	$n_q = 2$	$x^{(5)}$	0.83	0.67	0.98	$y(7)$	$y(6)$	$d^{(5)} = 0.79$

where the value $x_0 = -1$, associated with the neural threshold, must be inserted in all neurons of the network.

According to the previous table, at the first instant $\{t = 4\}$, the recurrent MLP will use only the time delay inputs, since the feedback of the last two values produced by the network are null because there is no past outputs. However, at the next instant $\{t = 5\}$, the first value produced by the network output $\{y(4)\}$, obtained in the previous instant, will then be available. Successively, for the posterior instant $\{t = 6\}$, both previous outputs $\{y(5), y(4)\}$ that have been already produced by the network at the two previous instants are then used as inputs. Such process repeats for all posterior instants in order to compile the training set of the recurrent MLP.

The prediction of future values related to the behavior of the process, which will be performed after the network is trained, will be similarly executed using those same values of n_p and n_q assumed during the learning process. In this example, assuming that the operation domain is $\{t \ge 9\}$, consider the following table:

	x_1	x_2	x_3	x_4	x_5	estimated output	
$t=9$	$x(8)$	$x(7)$	$x(6)$	$y(8)$	$y(7)$	$x(9) \approx y(9)$	
$t=10$	$x(9)$	$x(8)$	$x(7)$	$y(9)$	$y(8)$	$x(10) \approx y(10)$	
$t=11$	$x(10)$	$x(9)$	$x(8)$	$y(10)$	$y(9)$	$x(11) \approx y(11)$	
$(...)$	$(...)$	$(...)$	$(...)$	$(...)$	$(...)$	$(...)$	

Thus, for any future instant in the operation domain of the recurrent MLP, the prediction of its values will always take into account the three last time-delayed inputs and the last two outputs produced by the network.

This topological configuration, in which just the results produced by the output neurons are fed back to the network inputs, is also known as the Jordan neural network or single recurrent network (Jordan 1986). On the other hand, the configuration may also be called Elman neural network if only the outputs produced by the intermediate layers are fed back to the context unities (Elman 1990), thus producing semi-recurrent (local) signals.

The recurrent MLP can be used as an estimator of the future behaviors of a dynamic system and can be converted in several configurations with applicability in control systems. One of the first investigations in this theme was developed by Narendra and Parthasarathy (1990), and more detailed investigations can be found at Leondes (2006), Norgaard et al. (2010), Suykens et al. (2001).

5.5 Aspects of Topological Specifications for MLP Networks

The most suitable MLP topology to map a particular problem is usually done empirically, since the design of the network depends on the adopted learning algorithm, on how the weight matrices were initialized, on the complexity of the problem, on the spatial distribution of the training samples and, nonetheless, on the quality of the training set available. This last aspect can be substantially affected by the amount of noise in the samples, by the existence of errors, and by the existence of outliers.

As an example, consider four candidate topologies for mapping a given problem, all of them consisting of a single hidden layer:

- Candidate topology 1 → 05 neurons in the hidden layer.
- Candidate topology 2 → 10 neurons in the hidden layer.
- Candidate topology 3 → 15 neurons in the hidden layer.
- Candidate topology 4 → 20 neurons in the hidden layer.

The objective now is to define which one of these topologies would be the best suitable to map the problem.

The cross-validation method is one of the most used statistical methods for selecting the best MLP topology (Kohavi 1995; Ripley 1996). Its purpose is to evaluate the performance of each topology when using a data set different from that used in the training process. As it will be discussed in the next subsection, three cross-validation methods are usually employed in the process of selecting the MLP topology.

5.5.1 Aspects of Cross-Validation Methods

The first method is called random subsampling cross-validation, where the entire data set (samples) available is randomly divided into two subsets: training and test (validation).

More specifically, the training subset, as the name implies, will be used to train all candidate topologies, while the test subset is only applied for selecting the topology that provides better generalization results. In fact, as the samples from the test subset do not participate in the learning process, it is possible to evaluate the generalization performance of each candidate topology by comparing the results produced by their outputs with the desired values when using the test subset.

In practical terms, from the total data set available, around 60–90 % of its elements are chosen at random for the training subset. This partitioning system must be repeated several times during the learning process of the candidate topologies so to provide (for each trial) different samples in both subsets. The global performance

Total data set

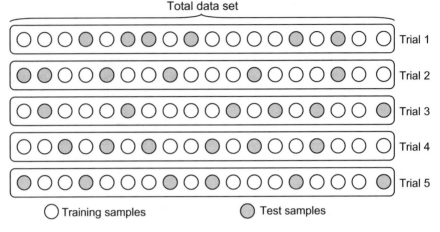

Fig. 5.34   Random subsampling cross-validation method

of each candidate topology will be obtained from the average of the individual performances in each trial.

Figure 5.34 illustrates the strategy for selecting samples for the training and test subsets. For each of the five trials, six of the eighteen samples available will be part of the test subset, while the remaining twelve will be part of the training subset. In each trial, the samples of the test subset will be randomly chosen considering all the available samples (without verifying whether some of them have been already chosen in previous trials or not).

The second cross-validation method used for designing MLP networks is known as k-fold cross-validation. It consists of dividing the total sample set in k partitions, where $(k - 1)$ partitions will be used to compose the training subset and the remaining partition will be used to compose the test subset.

After this division, the learning process is repeated k times until all partitions have been used as a test subset. Figure 5.35 exemplifies the strategy of this mechanism for 20 samples, with $k = 5$, which implies performing 5 trials.

The value of the parameter k is linked to the total number of available samples, and it is usually defined between 5 and 10. The global performance of each candidate topology is obtained by evaluating the average of the individual performances at each one of the k trials.

Finally, the third method (and least usual) is called leave-one-out cross-validation, which consists of using a single sample for the test subset, while the other samples are used in the training subset. The learning process is then repeated until all samples are individually used as the test subset.

This last technique is a particular case of the k-fold method when the parameter k assumes the value corresponding to the total number of samples. However, there is a high computational effort in this case, because the learning process will be

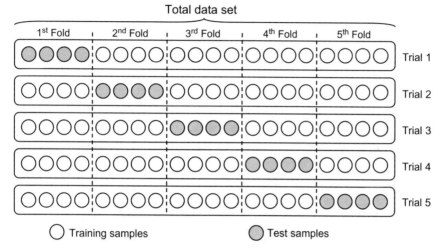

Fig. 5.35 k-fold cross-validation method

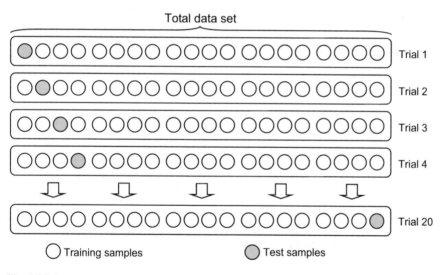

Fig. 5.36 Leave-one-out cross-validation method

repeated, for each candidate topology, as many times as the size of the complete sample set. Figure 5.36 summarizes this cross-validation strategy for 20 samples.

The algorithm steps for implementing any one of the three cross-validation methods is as follows:

Begin {CROSS-VALIDATION algorithm}

> <1> Define the candidate topologies for the given problem;
>
> <2> Acquire the training and test subsets;
>
> <3> Apply the MLP learning algorithm to all candidate topologies using the training subsets;
>
> <4> Apply the test subset to the (already trained) candidate topologies in order to evaluate their generalization potential;
>
> <5> Obtain the final performance metric of each candidate topology with respect to the number of trials;
>
> <6> Select the candidate topology that obtained the best global performance;
>
> <7> If the global performance of the best candidate topology is within the precision required by the problem,
>
> <7.1> then: End the cross-validation process.
>
> <7.2> else: Specify a new set of candidate topologies and go back to step <3>.

End {CROSS-VALIDATION algorithm}

In contrast to all the three methods, when all available samples are used both for learning and validation, it becomes impossible to verify the generalization capability of the network. What would be evaluated is its capability of satisfactorily memorize the desired responses with the respect to the presented patterns.

Alternatively, instead of using cross-validation methods, two rules are widely used for specifying a topology of an MLP constituted by a single hidden layer. These rules are given by:

$$n_1 = 2 \cdot n + 1 \quad \{\text{Kolmogorov method}\} \tag{5.65}$$

$$2 \cdot \sqrt{n} + n_2 \leq n_1 \leq 2 \cdot n + 1 \quad \{\text{Fletcher-Gloss method}\}, \tag{5.66}$$

where n is the number of inputs of the network, n_1 is the number of neurons in the hidden layer, and n_2 is the number of neurons in the output layer. In particular, for problems of pattern classification with n_c classes, another heuristics that have been usually used is that implemented by the WEKA platform (Waikato Environment for Knowledge Analysis) (Witten et al. 2011), that is:

$$n_1 = \frac{n + n_c}{2} \tag{5.67}$$

These networks, although largely used, are appropriate for just a few well-behaved classes of a problem, since they do not consider attributes that are, in fact, very relevant for specifying the topology of MLP networks. Such attributes are the quantity of data, the complexity of the problem, the quality of data and their disposition in the sampling space.

5.5.2 Aspects of the Training and Test Subsets

Some attention must be paid when compiling the training subset. It is necessary to ensure that all samples representing the maximum value and minimum value of each input variable belong to these subsets. Otherwise, if such values are allocated in the test subset, then the MLP could generate significant errors, since it would try to generalize values that are outside of the definition domain of its input variables.

In fact, one must have in mind that the MLP, when projected for problems of functional approximation or pattern recognition, will always disregard which is the process behavior outside its definition domains, because only samples from that definition domain were used during learning.

Consequently, during the whole operation stage, it must be guaranteed that the input signals of each input variable are, yet again, encompassed within the definition domains that were obtained from the maximum value and minimum value of the training subset.

To illustrate these project aspects, Fig. 5.37 shows a scenario where the MLP was trained to map the function sine(x), whose training samples belonged to the domain between 0 and 10.

By inspecting Fig. 5.37, it is possible to conclude that the MLP accurately estimated the values within the respective definition domain.

However, outside this domain, the estimation will result in inaccurate values. Certainly, as already stated in this chapter, such fact occurs because the MLP ignores the function behavior at both the left and right side of the definition domain.

5.5.3 Aspects of Overfitting and Underfitting Scenarios

It is important to highlight that increasing indiscriminately the number of neurons, as well as increasing the number of intermediate layers, does not ensure the appropriate generalization of the MLP with respect to the samples of the test subset.

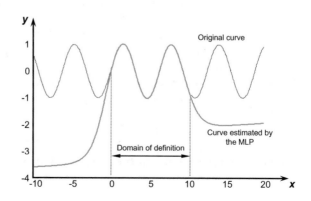

Fig. 5.37 MLP estimation considering points inside and outside the definition domain

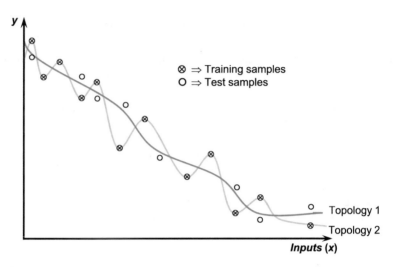

Fig. 5.38 Behavior of MLP networks operating with and without overfitting

Invariably, such premature actions usually drive the MLP to overfit, a condition in which the network memorizes the responses to the input stimuli. In this scenario, the squared error of the learning process tends to be very low; however, during the generalization stage, when the test subset is presented to the network, the squared error tends to be very high. This fact is known as the overfitting condition.

Figure 5.38 shows the behavior of an MLP operating with overfitting condition (topology 1), and, in contrast, the behavior of an MLP with good generalization (topology 2), that is, without overfitting.

It is possible to see in Fig. 5.38 that topology 2 (with overfitting), which contains 20 neurons, will certainly produce a small squared error during the training stage, since the training samples are almost coincident with the curve mapped by the network. However, when the test subset is used as the input to the network, topology 2 will generate a significant error, since the samples are distant from the curve produced. Using the terminology of the system identification area, it is possible to say that topology 2 is behaving like a biased estimator, whose values produced are very divergent from the expected (Ljung 1999).

On the other hand, topology 1 (without overfitting), constituted by 10 neurons, will provide lower error for the test samples, since its output curve represents more faithfully the behavior of the process. Thus, this network behaves as an unbiased estimator because its responses are within an acceptable error margin.

Another graphic that illustrates overfitting is shown in Fig. 5.39, in which the MLP is operating with an excessive memorization of the training samples of function sine(x), which is affected by noise.

In contrast, the curve produced by the MLP in Fig. 5.40 (without overfitting), when trained with the same samples used in Fig. 5.39, will probably produce better generalization responses, since its geometric shape closer to that expected for the

Fig. 5.39 MLP used for mapping the function sine(x), with overfitting

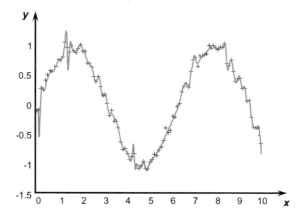

Fig. 5.40 MLP used for mapping the function sine(x), without overfitting

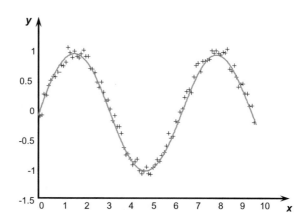

sine(x) function. Besides, this not overfitted MLP naturally eliminated noises, since its memorization was not excessive.

On the other hand, an MLP topology with a reduced number of neurons might be insufficient for extracting and storing features of the process. Therefore, the network will not be able to build hypotheses about the process behavior, which results in a situation of natural debility called underfitting. In this case, the squared error in both learning and test stages are very significant.

5.5.4 Aspects of Early Stopping

A simple procedure that can be used alongside with cross-validation methods is the early or anticipated stopping, in which the learning process of a candidate topology is continuously monitored, and the performance of the network is evaluated with respect to the samples of the test subset. The training process is stopped when the

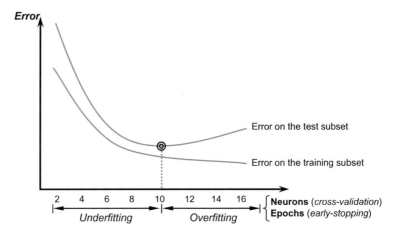

Fig. 5.41 Underfitting and overfitting examples

mean squared error begins to increase (when the network is evaluated with the test subset) between successive epochs.

In fact, that abrupt variation indicates the beginning of overfitting of the training subset, for example, when learning measurement noises (Finnof et al. 1993). In the example of Fig. 5.41, the learning process should end at the tenth training epoch.

In summary, the process of specifying the topology of an artificial neural network must take into account the tradeoff between overcome underfitting and avoid overfitting. To illustrate such procedure, consider an MLP with a single hidden layer, Fig. 5.41 illustrates the cases of underfitting and overfitting with respect to the number of neurons in the hidden layer.

By analyzing Fig. 5.41, it is possible to see that the most appropriated topology for that case would have a total of 10 neurons in its hidden layer. A number smaller than 10 would result in a significant error with respect to the test subset due to the insufficiency of neurons (underfitting). On the other hand, for a number greater than 10, the MLP would also begin to produce a considerable error with respect to the samples of the test subset, regardless that the squared error obtained in the training subset tends to be gradually lower when more neurons are added (overfitting).

As a last remark, it is important to note that the early stopping method must be individually applied to every candidate topology. However, the selection of the best candidate topology is still performed by cross-validation methods.

5.5.5   Aspects of Convergence to Local Minima

As the error surface produced by an MLP is nonlinear, the learning process could drive the weight matrices of the network in the direction of a local minimum, which may not correspond to the most satisfactory result.

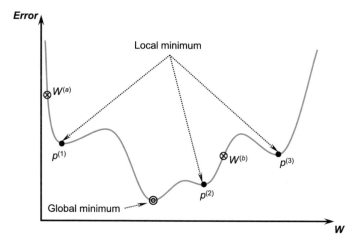

Fig. 5.42 Local minima points associated with the error function

Figure 5.42 illustrates some scenarios involving points of local minima. In that case, the network converges to the point indicated by $p^{(2)}$, and it will certainly produce better results than that from the case in which the network converges to $p^{(1)}$ or $p^{(3)}$.

The convergence to a given minimum point depends mostly on where the weight matrix W was initialized, since most learning algorithms are based on gradient descent methods. For example in Fig. 5.42, if the network had its weight matrix initiated in $W^{(a)}$, it would converge to the local minimum $p^{(1)}$. If it was initiated in $W^{(b)}$, it would likely go in the direction of $p^{(2)}$.

To avoid the network convergence to inappropriate local minima, one of the adopted practical procedures consists of executing the training process of each candidate topology more than once with different initial weights (generated at random). Thus, depending on the spatial position of its initial weight matrices, the network could converge to all local minima, or even to the global minimum, where the network would provide a better representation of the process behavior.

More sophisticated possibilities, such as those based on simulated annealing (Kirkpatrick et al. 1983), on tabu search (Glover and Laguna 1998), genetic algorithms (Goldberg 2002) and Monte Carlo method (Pincus 1970), can also be incorporated in learning algorithms as strategies for escaping of local minimas.

5.6 Implementation Aspects of Multilayer Perceptron Networks

This section presents some guiding aspects about the implementation of multilayer perceptron networks.

(a)   **(b)**

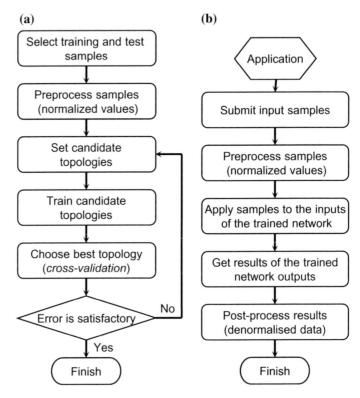

Fig. 5.43 Block diagram of the training and operation stages. **a** Training stage. **b** Operation stage

Figure 5.43 shows the block diagram that illustrates the main steps for implementing a network, considering both the training and operation stages. Each of those steps can be implemented using subroutines (functions and procedures), whose internal instructions can also be refined or grouped for enhancing the computational efficiency.

As shown in Fig. 5.43a, it is necessary to preprocess the training/test patterns in order to enhance the training performance. Such aspect usually implies the scaling of the patterns to the range of the activation function used in the hidden layers, typically represented by the logistic or hyperbolic tangent functions. This scaling is made to avoid the saturation of neurons (Fig. 5.9).

It is recommended to scale the input and output signals, for both training and test patterns, considering the dynamic ranges of the activation functions. One of the most used techniques is based on the proportional segment principle (Thales' Theorem) illustrated in Fig. 5.44. A set of values defined initially within a range $x \in [x^{min}, x^{max}]$ will be converted to a proportional range between -1 and 1, which can represent the dynamic ranges of the activation functions.

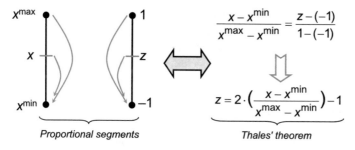

Fig. 5.44 Normalization principle of the training and test patterns

Thus, considering that the logistic activation function will be used by the neurons of the intermediate layer, Fig. 5.45 illustrates the normalization to be performed on the inputs and outputs of the training and test patterns.

Similarly, Fig. 5.46 describes the normalization to be performed when the hyperbolic tangent activation function is used by the neurons of the intermediate layers.

It is important to highlight that all input and output variables need to be individually normalized with respect to their maximum and minimum values, considering all the available data. The maximum and minimum values must be within the

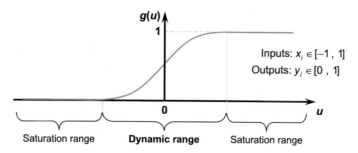

Fig. 5.45 Normalization domains for the logistic activation function

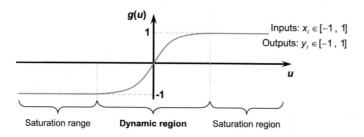

Fig. 5.46 Normalization domains for the hyperbolic tangent activation function

training set. Otherwise, these values will be on the test set, which implies a reduction of the domain of the training set.

A post-processing stage is required when the MLP is being used in the operation stage, as shown in Fig. 5.43b. In this case, operations of denormalization are made to convert the network responses to the values that represent the real application domains.

As it can be seen throughout this chapter, there are several parameters that can be adjusted to improve the performance of the MLP training process. Detailed investigations related to these various strategies are presented in Hagan et al. (2014), Reed and Marks II (1999).

Finally, also considering other relevant aspects involved in the training and operation stages of the MLP, the following practical remarks are presented:

(a) Increasing the number of neurons or layers of the MLP does not directly result in enhancements of its generalization ability.

(b) Considering two MLP topologies that result in the same generalization precision, always select the network composed of fewer neurons, since the network was able to extract more knowledge.

(c) Allocate to the training set all samples containing the minimum and maximum values with respect to any input or output variable of the MLP.

(d) Perform the training of each candidate topology several times, assuming random values for the weight matrices, in order to escape from regions of local minima.

(e) Initialize all weight matrices $W_{ji}^{(L)}$ with small random values. This recommendation comes from the analysis performed by LeCun (1989). The values must be within the range of $\left(-2.4/N \leq W_{ji}^{(L)} \leq 2.4/N\right)$, where N is the number of inputs of the MLP. The strategy tries to avoid that the weighted sum of the input values saturates the activation functions.

(f) Associate suitable values to the learning (η) and momentum (α) terms, which must be within the ranges of $(0.05 \leq \eta \leq 0.75)$ and $(0 \leq \alpha \leq 0.9)$. The practical examples in Sect. 5.8 suggest some typical values for these parameters.

(g) Impose a maximum number of epochs as an additional stop criterion for the MLP training because it is a simpler and efficient strategy for stopping the training when the specified precision $\{\varepsilon\}$ becomes unachievable.

(h) Adopt the method of Levenberg-Marquardt for speeding-up the MLP training since it was proven to be the fastest method in several tests. However, for problems of function approximation and time-variant systems, the Resilient-Propagation method is usually the fastest, because it directly uses the signals of the gradient derivatives, which is an important aspect of these problems, since the desired outputs may change abruptly.

(i) Normalize the input and output samples to avoid the saturation domains of activation functions.

(j) Adopt the hyperbolic tangent as activation function for the neurons of the hidden layers. Because of its antisymmetric nature (odd function), this function contributes to enhancing the convergence process of the network.
(k) Always use the data in the test subset for evaluating the generalization potential.
(l) Apply preprocessing techniques and/or tools for feature extraction (e.g., Fourier transform, Wavelet transform, principal component analysis, and so on) for minimizing redundancies and reducing the dimensional complexity of the input signals.

5.7   Exercises

1. Explain if it is possible to perform the training of an MLP network by using the backpropagation algorithm when all weight matrices are initialized with null values. Also, point out what are the implications, if any, when all elements of the weight matrices are initialized with the same value (not null).
2. For problems involving function approximation, explain if there is any advantage and/or disadvantage of using the linear activation function in the neurons of the output layer instead of using the hyperbolic tangent function.
3. Explain what is underfitting and overfitting, and describe what actions can be taken to detect and avoid them.
4. Discuss the eventual consequences of using very high values for the learning term $\{\eta\}$ and very low values for the momentum term $\{\alpha\}$.
5. Consider a pattern classification problem composed of only two inputs, and whose separability boundary is represented by a compact convex region (closed and limited). Make an estimate of the minimum number of neurons that could be associated with the first hidden neural layer.
6. For the previous exercise, consider the condition in which the separability boundary between classes is represented by two disjoint compact regions, where one of the regions is convex and the other non-convex. Make an estimate of the minimum number of neurons in both the first and second hidden layer.
7. Consider the application of an MLP in a function approximation problem. The network inputs are the temperature and pressure while the output provides the heat to be inserted into a boiler. The MLP (after the training is finished) will operate for assisting the boiler control. Explain what would be the first aspect to be verified in new temperature and pressure values that will be inserted in the inputs of the network.
8. Explain if it is possible to assume null values for the threshold $\{\theta\}$ of all neurons of the last layer of the MLP, when the network is applied to problems involving function approximation.

9. Discuss if it is possible to use an MLP in function approximation problems when the linear activation function is assumed for all neurons in its intermediate layer.

10. Discuss the need for preprocessing training and test samples, highlighting its influence on the speed of the MLP training.

5.8   Practical Work 1 (Function Approximation)

When implementing a magnetic resonance image processor, it was observed that the variable $\{Y\}$, which measures the absorbed energy of the system, could be estimated by measuring three other variables $\{x_1, x_2, x_3\}$. However, due to the problem complexity, it is known that this mapping is difficult to be made by using traditional methods, and that the mathematical model available produces unsatisfactory results.

Thus, the team of engineers and scientists intends to use a Multilayer Perceptron as a universal curve fitting tool, with the objective of estimating (after the training is complete) the absorbed energy $\{y\}$ with respect to the values of x_1, x_2, and x_3. The topology of the network, which is composed of two neural layers, is illustrated in Fig. 5.47.

Using the backpropagation algorithm (generalized delta rule) and the training samples found in Appendix C, then do the following activities:

1. Execute five training processes for the MLP network, initializing the weight matrices with appropriate values in each training procedure. If necessary, update the random number generator in each training process so that the initial elements are different on each routine. Use the logistic activation function (sigmoid) for all neurons, with the learning rate value $\{\eta\}$ equal to 0.1 and precision $\{\varepsilon\}$ equal to 10^{-6}. The training set is found in Appendix C.

2. Register the results from the five training routines in the following table

3. Consider, from Table 5.2, the training routines with the two higher number of epochs and plot the corresponding mean squared error of each training epoch. Print both graphs on the same page without overriding them.

Fig. 5.47 MLP topology (practical work 1)

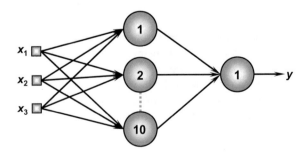

Table 5.2 Training results (practical work 1)

Training routine	Mean squared error	Total epoch number
#1 (T1)		
#2 (T2)		
#3 (T3)		
#4 (T4)		
#5 (T5)		

4. Based on Table 5.2, explain why both the mean squared error and the total number of epochs diverge according to the training process.
5. For all training processes executed in item 2, perform the validation of the network by using the test set provided in Table 5.3. Compute, for each training routine and for all test samples, the mean relative error (%) between the desired values and the values produced by the network. Also, compute the corresponding variance.
6. Based on Table 5.3, point out what final training configuration {T1, T2, T3, T4, or T5} would be the most suitable for the magnetic resonance system, that is, what network provides the better generalization.

Table 5.3 Set of test patterns (practical work 1)

Samples	x_1	x_2	x_3	d	$y(T1)$	$y(T2)$	$y(T3)$	$y(T4)$	$y(T5)$
1	0.0611	0.2860	0.7464	0.4831					
2	0.5102	0.7464	0.0860	0.5965					
3	0.0004	0.6916	0.5006	0.5318					
4	0.9430	0.4476	0.2648	0.6843					
5	0.1399	0.1610	0.2477	0.2872					
6	0.6423	0.3229	0.8567	0.7663					
7	0.6492	0.0007	0.6422	0.5666					
8	0.1818	0.5078	0.9046	0.6601					
9	0.7382	0.2647	0.1916	0.5427					
10	0.3879	0.1307	0.8656	0.5836					
11	0.1903	0.6523	0.7820	0.6950					
12	0.8401	0.4490	0.2719	0.6790					
13	0.0029	0.3264	0.2476	0.2956					
14	0.7088	0.9342	0.2763	0.7742					
15	0.1283	0.1882	0.7253	0.4662					
16	0.8882	0.3077	0.8931	0.8093					
17	0.2225	0.9182	0.7820	0.7581					
18	0.1957	0.8423	0.3085	0.5826					
19	0.9991	0.5914	0.3933	0.7938					
20	0.2299	0.1524	0.7353	0.5012					
Mean relative error (%)									
Variance (%)									

5.9   Practical Work 2 (Pattern Classification)

In a beverage processing, the use of a given preservative is a function of four variables defined as x_1 (water quantity), x_2 (acidity), x_3 (temperature), and x_4 (surface tension). There are three preservatives available that could be applied, which are defined by A, B and C. In sequence, laboratory experiments were carried out to specify which preservative must be mixed to a given beverage.

After 148 laboratory experiments, the team of engineers and scientists decided to use a Multilayer Perceptron network as a pattern classification tool for identifying what kind of preservative will be mixed with a given beverage. An MLP network with three outputs will be used, as presented in Fig. 5.48.

The output, which represents the type of preservative to be used, was standardized according to Table 5.4.

Using the training data presented in Appendix C, execute the training process of the MLP network (four inputs and three outputs). The network must classify, according to the measured values x_1, x_2, x_3, and x_4, which preservative must be mixed with a given beverage. Then, perform the following activities:

1. Execute the training process of the MLP network, by using the default back-propagation algorithm and initializing the weight matrices with appropriate random values. Use the logistic activation function (sigmoid) for all neurons, with the learning rate value $\{\eta\}$ equal to 0.1 and precision $\{\varepsilon\}$ equal to 10^{-6}.
2. After that, execute the training process for the MLP network, by using the backpropagation algorithm with momentum and initializing the weight matrices

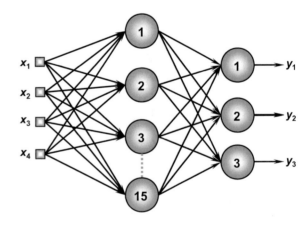

Fig. 5.48 MLP topology (practical work 2)

Preservative	y_1	y_2	y_3
A	1	0	0
B	0	1	0
C	0	0	1

Table 5.4 Standardization of MLP network outputs (practical work 2)

with the same values used in item 1. Use the logistic activation function (sigmoid) for all neurons, with the learning rate value $\{\eta\}$ equal to 0.1, momentum factor $\{\alpha\}$ equal to 0.9 and precision $\{\varepsilon\}$ equal to 10^{-6}.

3. For both training routines performed in item 1 and 2, plot the respective mean squared error $\{E_M\}$ in each training epoch. Print both graphs on the same page without overriding them. Measure also the processing time of each training routine.

4. Implement the routine that performs the post-processing of the outputs provided by the network (real values) to integer numbers. As a suggestion, adopt the symmetric rounding criterion, that is:

$$y_i^{\text{post}} = \begin{cases} 1, & \text{if } y_i \geq 0.5 \\ 0, & \text{if } y_i < 0.5 \end{cases} \qquad (5.68)$$

5. Perform the network validation using the test set provided in Table 5.5. Compute the mean relative error (%) between the desired values and the values produced by the network (after post-processing), for all test samples.

Table 5.5 Set of test patterns (practical work 2)

Samples	x_1	x_2	x_3	x_4	d_1	d_2	d_3	y_1^{post}	y_2^{post}	y_3^{post}
1	0.8622	0.7101	0.6236	0.7894	0	0	1			
2	0.2741	0.1552	0.1333	0.1516	1	0	0			
3	0.6772	0.8516	0.6543	0.7573	0	0	1			
4	0.2178	0.5039	0.6415	0.5039	0	1	0			
5	0.7260	0.7500	0.7007	0.4953	0	0	1			
6	0.2473	0.2941	0.4248	0.3087	1	0	0			
7	0.5682	0.5683	0.5054	0.4426	0	1	0			
8	0.6566	0.6715	0.4952	0.3951	0	1	0			
9	0.0705	0.4717	0.2921	0.2954	1	0	0			
10	0.1187	0.2568	0.3140	0.3037	1	0	0			
11	0.5673	0.7011	0.4083	0.5552	0	1	0			
12	0.3164	0.2251	0.3526	0.2560	1	0	0			
13	0.7884	0.9568	0.6825	0.6398	0	0	1			
14	0.9633	0.7850	0.6777	0.6059	0	0	1			
15	0.7739	0.8505	0.7934	0.6626	0	0	1			
16	0.4219	0.4136	0.1408	0.0940	1	0	0			
17	0.6616	0.4365	0.6597	0.8129	0	0	1			
18	0.7325	0.4761	0.3888	0.5683	0	1	0			
Total score (%)										

5.10   Practical Work 3 (Time-Variant Systems)

The price of given merchandise being traded in the financial market has a variation according to the data presented in Appendix C.

A research team is trying to apply artificial neural networks to predict the future behavior of this process. Thus, they proposed a MLP architecture, with the time delay neural network (TDNN) topology, as illustrated in Fig. 5.49.

The candidate topologies that can be applied to map this problem are given by:

TDNN-1 $\rightarrow$ 05 inputs (p = 05), with n_1 = 10.
TDNN-2 $\rightarrow$ 10 inputs (p = 10), with n_1 = 15.
TDNN-3 $\rightarrow$ 15 inputs (p = 15), with n_1 = 25.

By using the algorithm backpropagation with momentum and the training data presented in Appendix C, perform the following activities:

1. Execute 3 training routines for each candidate network, initializing the weight matrices (in each training process) with appropriate random values. If necessary, update the random number generator in each training process so that the initial elements are different on each routine. Use the logistic activation function (sigmoid) for all neurons, with the learning rate value $\{\eta\}$ equal to 0.1, momentum factor $\{\alpha\}$ equal to 0.8, and precision $\{\varepsilon\}$ equal to 0.5×10^{-6}.
2. Register the results of each training routine in Table 5.6.
3. For all training routines executed in item 2, perform the validation of candidate networks using the test set provided in Table 5.7. Compute for each training

Fig. 5.49 MLP topology (practical work 3)

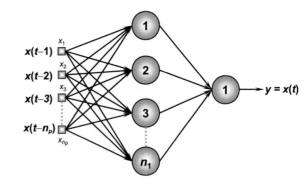

Table 5.6 Training results (practical work 3)

Training routine	TDNN-1		TDNN-2		TDNN-3	
	E_M	Epochs	E_M	Epochs	E_M	Epochs
#1 (T1)						
#2 (T2)						
#3 (T3)						

Table 5.7 Set of test patterns (practical work 3)

Values		TDNN-1			TDNN-2			TDNN-3		
Samples	$x(t)$	(T1)	(T2)	(T3)	(T1)	(T2)	(T3)	(T1)	(T2)	(T3)
$t = 101$	0.4173									
$t = 102$	0.0062									
$t = 103$	0.3387									
$t = 104$	0.1886									
$t = 105$	0.7418									
$t = 106$	0.3138									
$t = 107$	0.4466									
$t = 108$	0.0835									
$t = 109$	0.1930									
$t = 110$	0.3807									
$t = 111$	0.5438									
$t = 112$	0.5897									
$t = 113$	0.3536									
$t = 114$	0.2210									
$t = 115$	0.0631									
$t = 116$	0.4499									
$t = 117$	0.2564									
$t = 118$	0.7642									
$t = 119$	0.1411									
$t = 120$	0.3626									
Mean relative error (%)										
Variance (%)										

process the mean relative error between the desired values and the values produced by the network, for all test samples. Also, compute the respective variance.

4. For each candidate topology of Table 5.7, consider the best training routine {T1, T2, or T3} and plot the mean squared error (E_M) with respect to each training epoch. Print the three graphs on the same page without overriding them.

5. For each topology presented in Table 5.7, consider the best training routine {T1, T2, or T3} and plot the desired values and the values estimated by the network for the estimation domain ($t = 101,...,120$). Print the three graphs in the same page without overriding them.

6. Based on the analysis of the previous items, indicate what candidate topology {TDNN-1, TDNN-2, or TDNN-3} and what final training configuration {T1, T2, or T3} would be the most suitable for performing predictions in this process.

Chapter 6
Radial Basis Function Networks

6.1 Introduction

Radial Basis Function networks, commonly known as RBF, can also be employed in almost every kind of problems solved by MLPs, including those involving curve fitting and pattern classification.

Different from MLP networks, which can be composed of several intermediate layers, the RBF typical structure is composed of only one intermediate layer, in which the activation function is Gaussian, as illustrated by Fig. 6.1.

One of the main particularities of the RBF networks is the training strategy used for adjusting the weights of their both neural layers, which will be presented in detail in the next section. As shown in Fig. 6.1, another distinguishing feature of this network architecture is the activation function used by the neurons of the intermediate layer, which is always a radial basis function such as the Gaussian.

According to the classification presented in Chap. 2, RBF networks also belong to the multiple-layer feedforward architecture, whose training is supervised. From Fig. 6.1, it is possible to verify that the information that flows within its structure begins at the input layer, propagates to the intermediate layer (neurons with Gaussian activation function), and ends at the output neural layer (neurons with linear activation function).

6.2 Training Process of the RBF Network

The working principle of RBF networks is also similar to the principles of MLP networks, in which each input $\{x_i\}$, representing the signals from the application, will be propagated to the respective intermediate layer in the direction of the output layer.

However, different from the MLP, the training strategy of the RBF is composed of two very distinct steps or stages. The first stage, associated with adjusting the weights of the neurons in the intermediate layer, adopts a self-organized learning

© Springer International Publishing Switzerland 2017
I.N. da Silva et al., *Artificial Neural Networks*,
DOI 10.1007/978-3-319-43162-8_6

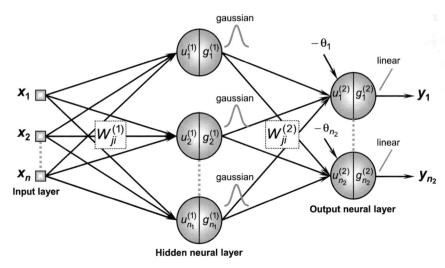

Fig. 6.1 Typical configuration of an RBF network

method (unsupervised), which depends only on the features of the input data. This adjustment is directly related to the allocation of the radial basis functions. On the other hand, the second stage, related to the weight adjustment of the neurons in the output layer, uses a learning criterion similar to the criterion employed in the last layer of the MLP, that is, the generalized delta rule.

Moreover, contrary to the MLP networks, the training process begins with the neurons from the intermediate layer and ends with the neurons of the output layer.

6.2.1   Adjustment of the Neurons from the Intermediate Layer (Stage I)

As mentioned before, the neurons belonging to the intermediate layer of the RBF are composed of activation functions with radial basis, being the Gaussian one of the most used. The expression that defines a Gaussian activation function is given by:

$$g(u) = e^{-\frac{(u-c)^2}{2\sigma^2}}, \tag{6.1}$$

where c defines the center of the Gaussian function and σ^2 denotes its variance (in which σ is equal to the standard deviation), which indicates how disperse is the activation potential $\{u\}$ in relation to the center $\{c\}$. Figure 6.2 illustrates the geometric shape of a typical Gaussian function.

As a comment, the higher the variance value, the larger is the base of the function. Figure 6.3 illustrates this features for a Gaussian function with three different values of variance, in which $\sigma_1^2 < \sigma_2^2 < \sigma_3^2$.

Fig. 6.2 Gaussian radial basis function

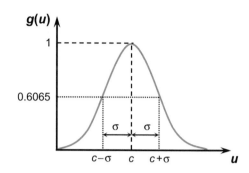

Fig. 6.3 Influence of the parameter σ^2 in the Gaussian function

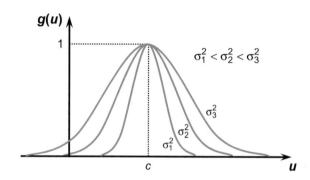

Thus, considering the expression provided by (6.1), the free parameters to be adjusted are the center c and the variance σ^2. In fact, taking into account the terminology adopted in Fig. 6.1, together with the configuration of the hidden neurons, the center c is directly associated with the weights of the hidden layer. In this situation, inputs $u_j^{(1)}$ are the input vector x, which represents the n external signals that are inputted in the network. Consequently, the output of each neuron j of the intermediate layer is expressed by:

$$g_j^{(1)}\left(u_j^{(1)}\right) = g_j^{(1)}(x) = e^{-\frac{\sum_{i=1}^{n}\left(x_i - w_{ji}^{(1)}\right)^2}{2\sigma_j^2}}, \text{ where } j = 1,\ldots,n_1 \qquad (6.2)$$

To illustrate the result above, Fig. 6.4 shows a Gaussian activation function with respect to two input signals x_1 and x_2, which composes a sample x.

It is possible to observe in Fig. 6.4 that as a sample (pattern) gets closer to the Gaussian center, the value produced by the radial domain of the activation function will become more significant (Powell 1987) and will tend to the value 1, as can be derived from (6.2). In such condition, the neuron will produce similar responses to all those patterns that are within a given radial distance from the Gaussian center.

Concerning problems of pattern classification, as discussed in Sect. 5.4.1, the MLP computes the delimiting boundaries between classes by combining

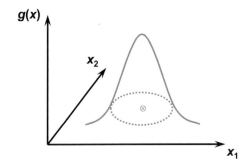

Fig. 6.4 Gaussian radial basis function

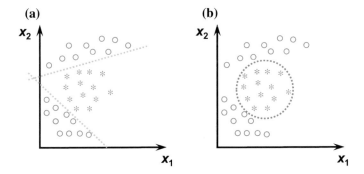

Fig. 6.5 The MLP and RBF decision boundaries. **a** MLP. **b** RBF

hyperplanes. In the case of the RBF with Gaussian activation function, the delimiting boundaries are defined by hyperspherical domains. Consequently, the pattern classification takes into account the radial distance of these patterns in relation to the center of these hyperspheres. Figure 6.5 illustrates these principles for a pattern classification problem composed of two inputs x_1 and x_2.

It is important to point out that these hyperspherical domains operate within the whole domain of the real numbers since this is also the domain of the Gaussian function. However, the most relevant values are those delimited by the variance and they are used to wrap samples of a given class. On the other hand, low activation values are produced for samples of the other class.

Also regarding Fig. 6.5, considering a two-dimensional problem, the decision boundary of the MLP is composed of straight line segments whereas the RBF delimits such boundary with a receptive domain that can be represented by a circle, as illustrated in Fig. 6.5b. Still in this hypothetical example, the MLP boundary is formed by two straight line segments, which indicates the existence of two neurons in the intermediate layer. In the case of the RBF, a single receptive domain can produce a boundary to group all the patterns of the same class, indicating the existence of only one neuron in the hidden layer.

Thus, the primary objective of the training process related to the neurons from the intermediate layer is to place the center of their Gaussian functions in an adequate point. One of the most used methods for this end is called the k-means method, whose purpose is to place the center of k-Gaussian functions in regions where the input patterns tend to group (Duda et al. 2001). It is noteworthy that the value of the parameter k is equal to the number of neurons in the intermediate layer, because the activation function of each one of the neurons is a Gaussian, as defined by (6.2), and the centers are represented by their corresponding weight vectors.

The sequence of computational procedures for the first training stage of RBF networks is shown through a pseudocode as follows:

BEGIN { RBF ALGORITHM – FIRST TRAINING PHASE}

<1> Obtain the set of training samples $\{x^{(k)}\}$;

<2> Initialize the weight vector of each neuron of the intermediate layer with the values of the n_1 first training samples;

<3> Repeat these instructions:

    <3.1> For all samples $\{x^{(k)}\}$, do:

        <3.1.1> Calculate the Euclidian distances between $x^{(k)}$ and $W_{ji}^{(1)}$, considering a single j-th neuron at each time;

        <3.1.2> Select the neuron j with the shortest distance in order to group the given sample with the closest center;

        <3.1.3> Attribute $x^{(k)}$ to group $\Omega^{(j)}$;

    <3.2> For all $W_{ji}^{(1)}$, with $j = 1..n_1$, do:

        <3.2.1> Adjust $W_{ji}^{(1)}$ according to the samples in $\Omega^{(j)}$:

$$W_{ji}^{(1)} = \frac{1}{m^{(j)}} \sum_{x^{(k)} \in \Omega^{(j)}} x^{(k)} \quad \{m^{(j)} \text{ is the number of samples in } \Omega^{(j)}\}$$

Until: there are no more changes in the groups $\Omega^{(j)}$ between two iterations;

<4> For all $W_{ji}^{(1)}$, with $j = 1..n_1$, do:

    <4.1> Calculate the variance of each Gaussian activation function by using the mean squared distance criterion:

$$\sigma_j^2 = \frac{1}{m^{(j)}} \sum_{x^{(k)} \in \Omega^{(j)}} \sum_{i=1}^{n} \left(x_i^{(k)} - W_{ji}^{(1)} \right)^2$$

END { RBF ALGORITHM – FIRST TRAINING PHASE}

To demonstrate the first training stage of an RBF and the k-means clustering method, consider a hypothetical problem with two inputs x_1 and x_2, whose training samples are represented in Fig. 6.6.

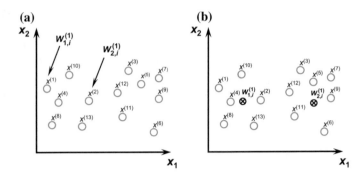

Fig. 6.6 Adjustment process using the *k*-means method. **a** Initial iteration. **b** Final iteration

Figure 6.6a shows the spatial placement of the samples used in the application of the *k*-means method. Assuming that, for this example, the RBF is composed of only two neurons in its intermediate layer ($n_1 = 2$). Thus, considering step <2> of the algorithm, it is possible to observe that the weight vectors of the first and second neurons receive, initially, the values assigned to $x^{(1)}$ and $x^{(2)}$, as also illustrated in Fig. 6.6a.

In sequence, step <3.1> consists of calculating the distances of each sample with respect to vectors $W_{1,i}^{(1)}$ (neuron 1) and $W_{2,i}^{(1)}$ (neuron 2). As an example, consider sample $x^{(3)}$. It is possible to observe in Fig. 6.6a that the sample is closer to $W_{2,i}^{(1)}$ than to $W_{1,i}^{(1)}$, while sample $x^{(4)}$ is clearly closer to $W_{1,i}^{(1)}$. In this case, sample $x^{(3)}$ is then inserted into the set $\Omega^{(2)}$ that contains all samples that activated neuron 2 as the winner (shortest distance), while sample $x^{(4)}$ is incorporated into set $\Omega^{(1)}$ that contains all samples that activated the neuron 1 as the winner. Repeating this procedure for all the other samples, the following sets $\Omega^{(1)}$ and $\Omega^{(2)}$ are then obtained:

$$\Omega_{\text{initial}}^{(1)} = \{x^{(1)}, x^{(4)}, x^{(8)}, x^{(10)}\}$$
$$\Omega_{\text{initial}}^{(2)} = \{x^{(2)}, x^{(3)}, x^{(5)}, x^{(6)}, x^{(7)}, x^{(9)}, x^{(11)}, x^{(12)}, x^{(13)}\}$$

In sequence, with the application of step <3.2>, the values of $W_{1,i}^{(1)}$ and $W_{2,i}^{(1)}$ are updated by taking into account all the samples from sets $\Omega^{(1)}$ and $\Omega^{(2)}$. Steps <3.1> and <3.2> are always iteratively repeated until there are no more changes in the sets $\Omega^{(1)}$ and $\Omega^{(2)}$ between two successive iterations. Thus, after achieving this convergence, vectors $W_{1,i}^{(1)}$ and $W_{2,i}^{(1)}$ are automatically positioned in the center of the groups, as illustrated in Fig. 6.6b, and the sets $\Omega^{(1)}$ and $\Omega^{(2)}$ have their final configurations given by:

$$\Omega_{\text{final}}^{(1)} = \{x^{(1)}, x^{(2)}, x^{(4)}, x^{(8)}, x^{(10)}, x^{(13)}\}$$
$$\Omega_{\text{final}}^{(2)} = \{x^{(3)}, x^{(5)}, x^{(6)}, x^{(7)}, x^{(9)}, x^{(11)}, x^{(12)}\}$$

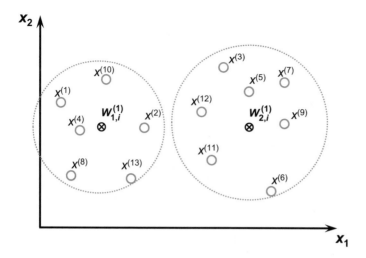

Fig. 6.7 Illustration of the receptive domains of the weight vectors

Finally, the receptive domains produced by the weight vectors of neurons 1 and 2, which are associated with each group, can be obtained by applying step <4.1>. Figure 6.7 illustrates the most-significant coverage radius of these receptive domains formed by the weight vectors adjusted by the k-means algorithm.

It is important to emphasize that the closer a given sample is to the center of the group, the greater is the significance of the response from the Gaussian receptive domain.

Several other clustering methods, such as those based on maximum-likelihood criterion (Nowlan 1989) or the recursive least mean square (Chen 1995; Chen et al. 1991), can also be used for the adjusting the weight vectors of the neurons from the intermediate layer.

6.2.2   Adjustment of Neurons of the Output Layer (Stage II)

The steps for adjusting the weights of the neurons in the output layer must be applied, only after the first training stage is finished.

The second training stage is performed by the same procedures used for training the output layer of the MLP, which are formulated in detail in Sect. 5.3.1. Thus, different from the first training stage of the RBF, this second stage uses a supervised learning method.

According to the terminology used in Fig. 6.1, the training set to adjust the free parameters related to the output neurons are composed of pairs of input and desired outputs, in which the inputs are the responses produced by the Gaussian functions of the neurons in the intermediate layer when fed with training samples, that is:

$$u_j^{(2)} = \sum_{i=1}^{n_1} \underbrace{W_{ji}^{(2)}}_{\text{parcel(i)}} \cdot \underbrace{g_i^{(1)}(u_i^{(1)})}_{\text{parcel(ii)}} - \theta_j, \quad \text{where } j = 1, \ldots, n_2, \qquad (6.3)$$

where $W_{ji}^{(2)}$ and θ_j are respectively the weights and thresholds from the output layer neurons, while the values from parcel (ii) are those computed with (6.2) using the values of $W_{ji}^{(1)}$ and σ_j^2 obtained on the first training stage.

Finally, similarly to the MLP, when a linear activation function defined by (1.11) is used in the output layer of the RBF, the output neurons only performs a linear combination of the Gaussian functions used by the neurons of the previous layer. Thus, from expression (6.3) and Fig. 6.1, the responses produced by the output neuron are given by:

$$y_j = g_j^{(2)}(u_j^{(2)}) = u_j^{(2)}, \quad \text{where } j = 1, \ldots, n_2 \qquad (6.4)$$

The instructions in pseudocode for the second training stage are explained as follows:

BEGIN {RBF ALGORITHM – SECOND TRAINING PHASE}

<1> Obtain the original training sample set {$x^{(k)}$};

<2> Obtain the desired output vector {$d^{(k)}$} for each sample;

<3> Initialize $W_{ji}^{(2)}$ with small random values;

<4> Specify the learning rate {η} and required precision {ε};

<5> For all samples {$x^{(k)}$}, do:

    <5.1> Obtain the values of $g_j^{(1)}$ with respect to the sample

        $x^{(k)}$;   {according to (6.2)}

    <5.2> Assume $z^{(k)} = [g_1^{(1)} \; g_2^{(1)} \ldots g_{n_1}^{(1)}]^T$;   {pseudo-samples}

<6> Initialize the epoch counter {epoch ← 0};

<7> Repeat the instructions:

    <7.1> $E_M^{previous} \leftarrow E_M$;   {according to (5.8)}

    <7.2> For all training pairs {$z^{(k)}$, $d^{(k)}$}, do:

      { Adjust $W_{ji}^{(2)}$ and θ_j by applying the same steps used on the adjustment of the synaptic weights of the MLP output layer (Subsection 5.3.1) }

    <7.3> $E_M^{current} \leftarrow E_M$;   {according to (5.8)}

    <7.4> epoch ← epoch + 1;

    Until: $\left| E_M^{current} - E_M^{previous} \right| \leq \varepsilon$

END {RBF ALGORITHM – SECOND TRAINING PHASE}

The variable *epoch* can also be used as a stop criterion for the second training stage, encompassing those scenarios when the specified precision for a given problem becomes unreachable.

Thus, the application of the first training stage (unsupervised), followed by the second stage (supervised), allows the adjustment of all the free parameters $(W_{ji}^{(1)}, \sigma_j^2, W_{ji}^{(2)}, \theta_j)$ of the RBF network. Additionally, the number of neurons in the intermediate layer (Gaussian activation functions) can also be stipulated by using techniques of cross-validation presented in Sect. 5.5.1.

Therefore, after performing both training stages, the RBF network can estimate the outputs of the process mapped when new samples are presented to its inputs. The sequence of steps for this operation is presented as follows:

BEGIN {RBF ALGORITHM – OPERATION PHASE}

<1> Present a sample $\{x\}$;

<2> Assume that the parameters $W_{ji}^{(1)}, \sigma_j^2, W_{ji}^{(2)}$ and θ_j are already adjusted during the training phases;

<3> Perform the following instructions:

<3.1> Obtain $g_j^{(1)}$;   {according to (6.2)}

<3.2> Obtain $u_j^{(2)}$;   {according to (6.3)}

<3.3> Obtain y_j;   {according to (6.4)}

<4> Provide the network outputs based on the values contained in y_j.

END {RBF ALGORITHM – OPERATION PHASE}

During the operation phase, step <3.3> can also be suppressed if the linear activation function (1.11) is adopted for the neurons in the output layer, since in such condition y_j is equal to the result of step <3.2>.

6.3 Applications of RBF Networks

RBF networks have been widely applied to solve problems involving curve fitting (system identification) and pattern classification, although there are many kinds of applications in which these networks have been successfully implemented.

Regarding curve fitting problems, deductions inspired by the studies of Kolmogorov (1957) are also valid for the RBF networks. To this end, Park and Sandberg (1991) demonstrated that such networks are considered universal curve

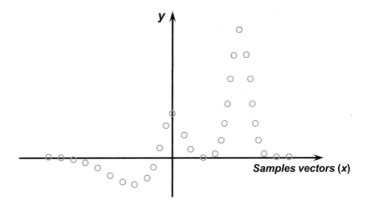

Fig. 6.8 Training samples for the curve fitting task using the RBF network

fitting tools, just like the MLP. In fact, by replacing the expression (6.3) in (6.4), the following relationship is obtained:

$$y_j = \sum_{i=1}^{n_1} \underbrace{W_{ji}^{(2)}}_{\text{parcel(i)}} \cdot \underbrace{g_i^{(1)}(u_i^{(1)})}_{\text{parcel(ii)}} - \theta_j, \quad \text{where } j = 1, \ldots, n_2 \qquad (6.5)$$

By analyzing Fig. 6.1, it is possible to conclude that a function y_j that is being mapped by the RBF is composed of a superposition of Gaussian activation functions {parcel (ii)}, which are represented by the terms $g_i^{(1)}(u_i^{(1)})$. These terms are linearly weighted by $W_{ji}^{(2)}$ {parcel (i)}, which is associated to the output layer.

To illustrate the curve fitting process, as also illustrated for the MLP case in Fig. 5.26, consider an RBF that implements a function approximation. Some training samples are shown in Fig. 6.8.

Figure 6.9 illustrates the RBF used in the fitting task, whose hidden layer is composed of three neurons.

After performing all training stages of the RBF, a possible final configuration for the linear combination of the Gaussian activation functions is shown in Fig. 6.10. These Gaussian activation functions are the outputs of neurons A, B, and C of the intermediate layer. In summary, the output neuron Y of the RBF produces a function that can map the behavior of the process presented in Fig. 6.8.

Through the examination of Fig. 6.10, it is possible to observe that the weight vectors of the neurons from the intermediate layer $\{W_{ji}^{(1)}\}$ are the responsible for translating the Gaussian activation functions into their domains. Indeed, these weights are representing the centers of the Gaussian functions, while the weights of the output layer $\{W_{ji}^{(2)}\}$ are responsible for their scaling.

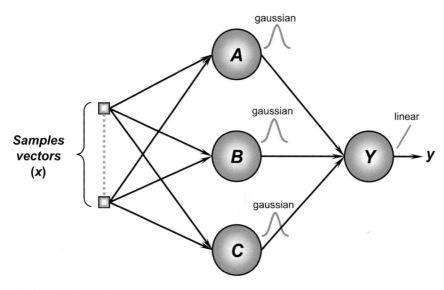

Fig. 6.9 Topology of the RBF used for curve fitting (function approximation)

An RBF, as well as an MLP, can be used to map any nonlinear continuous function defined on a compact (closed) domain. However, the enough number of neurons in the hidden layer for performing such task is still a current object of study. Consequently, cross-validation techniques, presented in Sect. 5.5.1, can also be used in RBF networks as a strategy for selecting the most suitable topology.

Regarding problems of pattern classification, the decision boundaries created by the Gaussian activation functions are formed by receptive hyperspherical domains, whose coverage radiuses produce more significant values for those points (samples) that are closer to the centers. As an example, for a hypothetical pattern classification problem, consider an RBF composed of two inputs $\{x_1$ and $x_2\}$, two neurons in its intermediate layer, and a single neuron in its output layer, as presented by Fig. 6.11.

Taking into account the first training stage of the RBF network, Fig. 6.12 illustrates a possible final configuration of the Gaussian functions associated with neurons A and B. The results are produced by the application of the k-means algorithm.

By analyzing Fig. 6.12, it is possible to verify that the Gaussian function produced by neuron A is larger (higher variance) than the one computed by neuron B.

Finally, for this illustrative example, Fig. 6.13 shows the neuron output Y, whose objective is to perform a linear combination of both Gaussian functions implemented by neurons A and B, in order to perform a correct pattern classification.

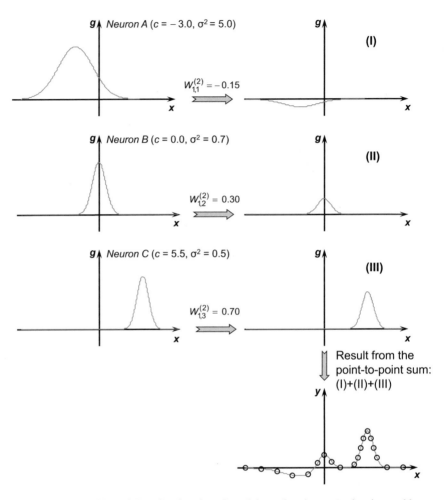

Fig. 6.10 Superposition of Gaussian functions for solving a function approximation problem

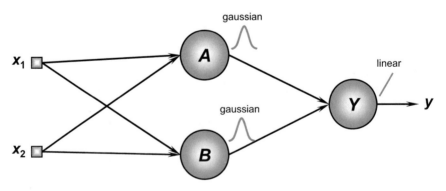

Fig. 6.11 Example of an RBF applied to a pattern classification problem

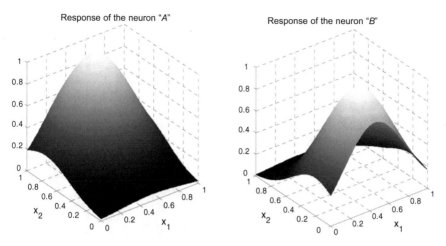

Fig. 6.12 Gaussian functions associated with neurons A and B

Figure 6.13 shows that neuron C inverted the Gaussian function produced by the output of neuron A, with the objective of correctly classifying the classes of the problem. For such, the value of the weight associating neuron A to neuron Y is negative, while the value of the weight connecting neuron B to Y is positive.

Finally, Fig. 6.14 illustrates the receptive domain, with hyperspherical (circle in 2D) shape, provided by the output neuron of the RBF. Differently from those boundaries that would be produced by the MLP, which are delimited by hyperplanes (2D straight lines), Fig. 6.14 shows a radial boundary resulting from the linear combination of the outputs produced by neurons A and B.

If the pattern recognition problem had three inputs, the receptive domain would be represented by a sphere and, thus, would be represented by hyperspheres for four or more input signals.

To exemplify the differences between MLP and RBF networks when applied to the same pattern classification problem, consider two classes of samples that must be classified based on the input signals given by x_1 and x_2, as illustrated by Fig. 6.15.

Figure 6.16 shows the geometry of the boundaries of a possible classification after training both networks. From the figure, it is possible to understand that the MLP intermediate layer could be composed of eight neurons, as eight straight lines are seen in the figure; while the RBF intermediate layer could be composed of only two neurons, which are characterized by the circles that delimit the boundaries.

Thus, it is important to emphasize that the classification boundaries are always defined by hyperplanes when using an MLP, and by hyperspherical receptive domains when using an RBF. Figure 6.17 illustrates a three-dimensional decision boundary based on the results from Fig. 6.16.

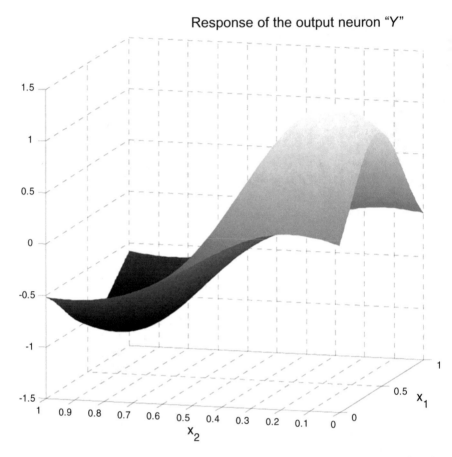

Fig. 6.13 Output of the neuron Y, which performs the linear combination of the two Gaussian functions produced by neurons A and B

In general, when much training samples are available, regardless of the problem being mapped, RBF networks may require more neurons on its intermediate layers when compared to MLP networks. However, the training of RBF networks is usually much faster when compared with MLP networks (Cowan et al. 1991).

6.4   Exercises

1. Considering the features of neural architectures, discuss three differences between MLP and RBF networks.
2. Explain what would be the eventual implication of assuming the same variance to the activation function of all neurons in the intermediate layer of RBF networks.

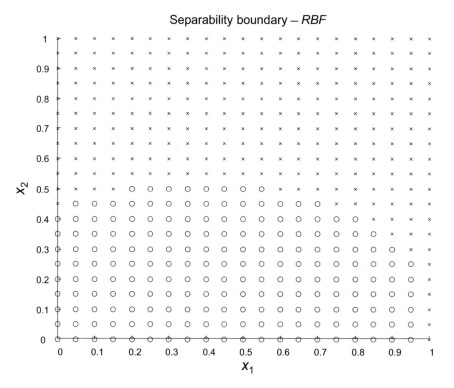

Fig. 6.14 Radial classification boundary for the RBF network

Fig. 6.15 Disposition of samples for pattern classification using MLP and RBF networks

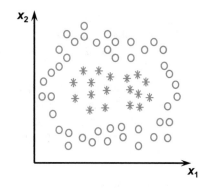

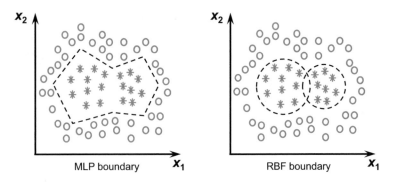

Fig. 6.16 Classification boundaries for the MLP and RBF

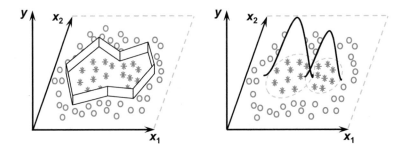

Fig. 6.17 Decision boundaries produced by the MLP and RBF networks

3. Discuss the need for using the activation threshold $\{\theta_j\}$ in all the neurons in the output layer of RBF networks.
4. Describe what is the purpose of the hyperspherical receptive domains produced by the Gaussian activation functions of neurons in the intermediate layer of RBF networks.
5. Write in pseudocode the steps required for adjusting $W_{ji}^{(2)}$ and θ_j during the second training stage of RBF networks.
6. Consider the exclusive-or (*Xor* gate) problem. Discuss the minimum number of neurons that would be required in the intermediate layer of the RBF for solving this problem.
7. Show that, in vector notation, the expression given by (6.2) is equivalent to the following equation:

$$g_j^{(1)}(x) = e^{\frac{\left(x - w_{(j)}^{(1)}\right)^T \cdot \left(x - w_{(j)}^{(1)}\right)}{2\sigma_j^2}},$$

where $W_{(j)}^{(1)}$ is the weight vector associated with the jth neuron of the intermediate layer.

8. Explain what are the required modifications in the vector expression of the previous exercise to produce hyperellipsoidal receptive domains.
9. Again, considering the exclusive-or problem, explain if the minimum number of neurons in the intermediate layer of the RBF would also be the same when using activation functions that produce hyperellipsoidal domains instead of radial.
10. For pattern classification problems, what would be the eventual advantages and disadvantage of RBF when compared to MLP.

6.5   Practical Work 1 (Pattern Classification)

To verify the existence of radiation in certain nuclear substances, concentration analysis of two variables $\{x_1$ and $x_2\}$ can be performed. It was decided to use a RBF for executing the pattern classification from 50 known situations. The network topology is illustrated by Fig. 6.18.

The standard output, representing the existence or inexistence of radiation, is defined by the terminology shown in Table 6.1.

Using the training data presented in Appendix D, train an RBF (2 inputs, 1 output) that can classify, with respect to the measurements of x_1 and x_2, if a given substance presents signs of radiation or not. In sequence, perform the following activities:

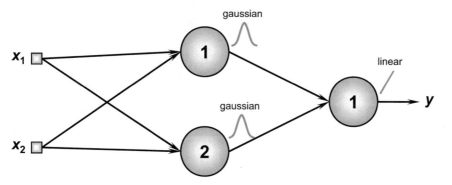

Fig. 6.18 RBF Architecture (practical work 1)

Table 6.1 RBF Standard output

Radiation status	Output (y)
Existent	1
Inexistent	-1

Table 6.2 Results of the first training stage

Cluster	Center	Variance
1		
2		

Table 6.3 Results of the second training stage

Parameter	Value
$W_{1,1}^{(2)}$	
$W_{2,1}^{(2)}$	
θ_1	

1. Train the hidden layer using the k-means algorithm. Since this is a pattern classification problem, compute the center of the two groups (clusters), taking into account only those patterns with radiation. After the training (first stage), provide the coordinates of the center of each cluster and its respective variance (Table 6.2).
2. Perform the second training stage using the generalized delta rule, with learning rate of 0.01 and precision of 10^{-7}. After meeting the convergence, provide the values of the weights and thresholds of the neurons in the output layer (Table 6.3).
3. Since this problem is a typical pattern classification process, implement the procedure that performs the post-processing of the responses computed by the network (real values) in order to produce the standard integer numbers presented in Table 6.1. For such, use the following function:

$$y^{\text{post}} = \begin{cases} 1, & \text{if } y \geq 0 \\ -1, & \text{if } y < 0 \end{cases},$$

where y^{post} provides the final result for the pattern classification problem, and is only used for post-processing the test set.

Table 6.4 Validation results (practical work 1)

Sample	x_1	x_2	d	y	y^{post}
1	0.8705	0.9329	−1		
2	0.0388	0.2703	1		
3	0.8236	0.4458	−1		
4	0.7075	0.1502	1		
5	0.9587	0.8663	−1		
6	0.6115	0.9365	−1		
7	0.3534	0.3646	1		
8	0.3268	0.2766	1		
9	0.6129	0.4518	−1		
10	0.9948	0.4962	−1		
Success rate (%):					

4. Validate the network by applying the test set shown in Table 6.4. Provide the success ratio (hite rate)(%) between the desired values and the responses produced by the RBF (after post-processing) for all test patterns.
5. Explain what strategies can be implemented to increase the hit rate of this RBF.

6.6   Practical Work 2 (Function Approximation)

The amount of gasoline $\{y\}$ that is fed into automobiles by an electronic fuel injection system can be computed in real-time using three measurements $\{x_1, x_2,$ and $x_3\}$. Given the inherent complexity of the process, which is classified as a nonlinear system, an artificial neural network is proposed to map the relationship between its inputs and outputs.

It is known that to compute the aforementioned mapping, which consists of a curve fitting problem (function approximation problem), two potential architectures can be applied, the MLP or the RBF. Since the engineering and scientific team has already performed the mapping using the MLP architecture, the objective now is to train an RBF so that the results provided from both networks could be compared.

For that purpose, perform the training of one RBF in order to compute the amount of gasoline $\{y\}$ that is fed to a motor, with respect to the variables x_1, x_2 and x_3. The topology of the RBF is illustrated on Fig. 6.19.

The configurations of different candidates, which may be applied to this problem, are specified as follow:

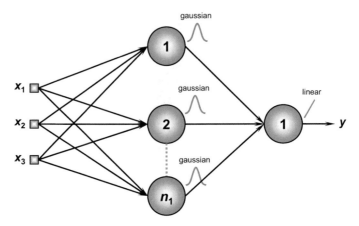

Fig. 6.19 RBF architecture (practical work 2)

Table 6.5 Results of the mean squared error (practical work 2)

Training	RBF-1		RBF-2		RBF-3	
	E_M	Epochs	E_M	Epochs	E_M	Epochs
#1 (T1)						
#2 (T2)						
#3 (T3)						

RBF-1 $\rightarrow$ Topology with $n_1 = 05$.
RBF-2 $\rightarrow$ Topology with $n_2 = 10$.
RBF-3 $\rightarrow$ Topology with $n_3 = 15$.

Using the training data presented in Appendix D, train the candidate topologies and perform the following exercises:

1. Perform 3 training processes for each RBF topology, initializing the weight matrix of the output layer with random values between 0 and 1. If necessary, restart the random number generator in each training process, so that the elements of the initial weight matrix are not the same. Use the learning rate $\{\eta\}$ equal to 0.01 and precision $\{\varepsilon\}$ equal to 10^{-7}.
2. Write the results of these three trainings $\{T1, T2, \text{ and } T3\}$ procedures in Table 6.5, considering all three candidates topologies.
3. For all training processes performed on item 2, validate the network with the values presented in Table 6.6. Provide for each training process the mean relative error (%) between the desired values and those estimated by the network. Also, obtain the respective variance (%).

Table 6.6 Results for validation (practical work 2)

Sample	x_1	x_2	x_3	d	RBF-1			RBF-2			RBF-3		
					$y(T1)$	$y(T2)$	$y(T3)$	$y(T1)$	$y(T2)$	$y(T3)$	$y(T1)$	$y(T2)$	$y(T3)$
1	0.5102	0.7464	0.0860	0.5965									
2	0.8401	0.4490	0.2719	0.6790									
3	0.1283	0.1882	0.7253	0.4662									
4	0.2299	0.1524	0.7353	0.5012									
5	0.3209	0.6229	0.5233	0.6810									
6	0.8203	0.0682	0.4260	0.5643									
7	0.3471	0.8889	0.1564	0.5875									
8	0.5762	0.8292	0.4116	0.7853									
9	0.9053	0.6245	0.5264	0.8506									
10	0.8149	0.0396	0.6227	0.6165									
11	0.1016	0.6382	0.3173	0.4957									
12	0.9108	0.2139	0.4641	0.6625									
13	0.2245	0.0971	0.6136	0.4402									
14	0.6423	0.3229	0.8567	0.7663									
15	0.5252	0.6529	0.5729	0.7893									
Mean relative error (%)													
Variance (%)													

4. For each topology presented in Table 6.6, consider the best training {T1, T2, and T3} performed on each one of them and plot the mean squared error (E_M) with respect to the training epochs. Print the three graphs on the same page without overlapping.

5. Based on the analysis of the previous items, indicate which candidate topology {RBF-1, RBF-2, and RBF-3} was the best and what final training {T1, T2, and T3} would be the most suitable for this problem.

Chapter 7
Recurrent Hopfield Networks

7.1 Introduction

As mentioned in Sect. 2.2.3, recurrent neural networks are those which the outputs of a neural layer can be fed back to the network inputs.

The best example of recurrent networks can be assigned, of course, to the one devised by Hopfield (1982), which are most commonly known as the Hopfield network. This neural network architecture, with global feedback, has the following characteristics:

- Dynamical behavior.
- Ability to memorize relationships.
- Possibility of storing information.
- Easy implementation in analogic hardware.

The work developed by Hopfield also contributed to trigger, at that time, a renewed and increased interest in artificial neural networks, which collaborated to the revival of important researches in the area, that were, somehow, stagnant since the publication of the book *Perceptron* by Minsky and Papert (1969).

Indeed, Hopfield's proposal addressed the existing links between recurrent neural architectures, dynamical systems and statistical physics, therefore, boosting curiosity of other areas of knowledge. His great triumph was to formulate several aspects that showed that recurrent neural networks with a single layer could be characterized by an energy function, which is related to the states of their dynamical behavior. Such architectures were also called as Ising models, a term used in analogy to ferromagnetism (Amit et al. 1985).

Given the above background, the minimization of the energy function $\{E(x)\}$ would take the network output to stable equilibrium points, which could be the desired solution for a particular problem. Figure 7.1 shows an illustration of stable and unstable equilibrium points.

© Springer International Publishing Switzerland 2017

I.N. da Silva et al., *Artificial Neural Networks*,

DOI 10.1007/978-3-319-43162-8_7

Fig. 7.1 Illustration showing stable and unstable equilibrium points

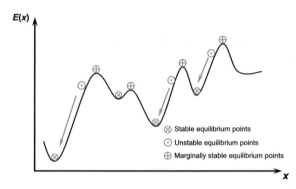

It can be noticed that an energy function can have several equilibrium considered stable. During the network convergence process, from some initial states, it is a tendency that these states always move toward one of the stable points (fixed points).

Besides the remarkable associative memories (Sect. 7.4), the main applications related to Hopfield networks are concentrated in the area of system optimization, such as dynamic programming (Silva et al. 2001; Wang 2004), linear programming (Malek and Yari 2005; Tank and Hopfield 1986), nonlinear constrained optimization (Silva et al. 2007; Xia and Wang 2004), and combinatorial optimization (Atencia et al. 2005; Hopfield and Tank 1985).

7.2 Operating Principles of the Hopfield Network

According to Fig. 7.2, the originally proposed Hopfield network is constituted of a single layer, in which all neurons are completely connected, this is, all network neurons are connected to all others and to itself (all network outputs are fed back to all network inputs).

Since the Hopfield network is composed of a single layer of neurons, the same terminology usually adopted in the literature is used to facilitate the understanding of the dynamics of the network. In a simplified way, the expression that dictates the behavior of each neuron in the Hopfield network, in continuous time, is given by:

$$\begin{cases} \dot{u}_j(t) = -\eta \cdot u_j(t) + \sum_{i=1}^{n} W_{ji} \cdot v_i(t) + i_j^b, & \text{with } j = 1, \ldots, n \quad (7.1) \\ v_j(t) = g(u_j(t)), & (7.2) \end{cases}$$

where:

$\dot{u}_j(t)$ is the inner state of the jth neuron, with $\dot{u}(t) = du/dt$;

$v_j(t)$ is the output of the jth neuron;

W_{ji} is the value of the synaptic weight connecting the jth neuron to the ith neuron;

Fig. 7.2 Conventional
Hopfield network

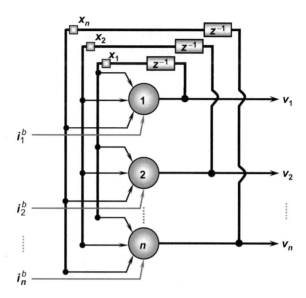

i_j^b is the bias applied to the jth neuron;

$g(.)$ is an activation function, monotonically increasing, that limits neuron outputs to a predetermined interval;

$\eta \cdot u_j(t)$ is a passive decay term; and

$x_j(t)$ is the jth input of each neuron (Fig. 7.2).

As a consequence, based on the interpretation of expressions (7.1) and (7.2), it is verified that the behavior of the Hopfield network is always dynamical and can be synthesized by the following steps:

(i)   Apply a set of signals $\{x\}$ in the inputs.
(ii) Obtain a vector $\{v\}$ of network outputs.
(iii) Feedback the inputs with the preceding outputs $\{x \leftarrow v\}$.
(iv) If the output vector $\{v\}$ remains stable (constant) between successive iterations, then, the network has converged to the final response (stable equilibrium point); otherwise, repeat steps (i) to (iii) until convergence.

To further clarify the steps involved in the dynamics of the Hopfield network, it is possible to interpret expressions (7.1) and (7.2) as follows:

- In $t = t_0 \Rightarrow$ input $x(t_0)$ generates output $v(t_0)$.
- In $t = t_1 \Rightarrow$ input $x(t_1) = v(t_0)$ generates output $v(t_1)$.
- In $t = t_2 \Rightarrow$ input $x(t_2) = v(t_1)$ generates output $v(t_2)$.
  (...)
- In $t = t_m \Rightarrow v(t_m) = v(t_{m-1})$, stable network condition.

Consequently, this sequence of successive iterations produces changes (increasingly smaller) in the network outputs, until their values become constant (stable). Depending on the configuration of the free parameters of the network, this process can be repeated indefinitely, and, therefore, generate unstable evolutions. Most networks with unstable dynamics present a chaotic behavior (Alligood et al. 2009), which has been an inspiration for several works related to Hopfield networks and chaotic systems, such as those investigated in Huang and Yang (2006) and Zhang and Xu (2007).

In most practical applications, where Hopfield networks were implemented by iterative computational algorithms, the discrete time model was used, in which expressions (7.1) and (7.2) are then converted to:

$$
\begin{cases}
u_j(k) = \sum_{i=1}^{n} W_{ji} \cdot v_i(k-1) + i_j^b, \text{ with } j = 1, \ldots, n & (7.3) \\
v_j(k) = g(u_j(k)), & (7.4)
\end{cases}
$$

where k is the iteration step.

As in the continuous case, given any initial condition $x^{(0)}$ and imposing some appropriated constraints on the weight matrix W, in order to guarantee stability (convergence to stable equilibrium points), the discrete time version always converges to those points corresponding to the solution of the problem.

7.3   Stability Conditions of the Hopfield Network

Due to their dynamical behavior, a vast domain of complex nonlinear systems can be represented by Hopfield networks. Depending on how the network parameters are chosen, the network could function as a stable system or as an oscillator, or as an entirely chaotic system (Aihara et al. 1990).

Most applications that comprise the use of the Hopfield Network require that the same behave as a stable system with multiple stable equilibrium points. The conditions that guarantee this type of behavior will be exposed in this section.

To analyze the evolution and stability of the Hopfield network, it is necessary to define an energy function or Lyapunov function associated with the network dynamics (Vidyasagar 2002). In other words, one has to prove that the system loses energy over time when considering some particular conditions imposed on it. Therefore, according to Lyapunov's second method, it must be shown that time derivatives of this function are always smaller or equal to zero. For this purpose, a Lyapunov function for the Hopfield network, whose neurons are changed asynchronously (one at a time) is defined by (Hopfield 1984):

$$
E(t) = -\frac{1}{2} v(t)^T \cdot W \cdot v(t) - v(t)^T \cdot i^b \tag{7.5}
$$

The mathematical expression for its time derivatives is obtained from (7.5), this is:

$$\dot{E}(t) = \frac{dE(t)}{dt} = (\nabla_v E(t))^T \cdot \dot{v}(t), \tag{7.6}$$

where ∇_v is the gradient operator with respect to vector v.

By imposing the first condition, which requires a symmetrical weight matrix $\{W = W^T\}$, the following relationship is obtained from (7.6) and (7.5):

$$\nabla_v E(t) = -W \cdot v(t) - i^b \tag{7.7}$$

By analyzing expression (7.1) and assuming that the passive decay term is null, the following relationship can be established from (7.7):

$$\nabla_v E(t) = -\dot{u}(t) \tag{7.8}$$

Therefore, substituting (7.8) into (7.6), it is obtained:

$$\begin{aligned} \dot{E}(t) &= -\dot{u}(t)^T \cdot \dot{v}(t) \\ &= -\sum_{j=1}^{n} \dot{u}_j(t) \cdot \dot{v}_j(t) = -\sum_{j=1}^{n} \dot{u}_j(t) \cdot \frac{\partial v_j}{\partial u_j} \cdot \frac{\partial u_j}{\partial t} \\ &= -\sum_{j=1}^{n} \underbrace{(\dot{u}_j(t))^2}_{\text{parcel(i)}} \cdot \underbrace{\frac{\partial v_j(t)}{\partial u_j(t)}}_{\text{parcel(ii)}} \end{aligned} \tag{7.9}$$

To finish this demonstration, so to extract the essential conditions for the Hopfield network stability, one should simply show that the time derivatives in expression (7.9) are always equal or less than zero. As the signal in the expression is negative, it should be shown that both portion (i) and portion (ii) produce positive values.

In fact, the parcel (i) always produces positive values, regardless of its argument, because it is a squared term. Finally, one must then show that the signal on parcel (ii) is also always positive. This condition is invariably true if activation functions that are monotonically increasing are used in the network, such as the logistic function and the hyperbolic tangent function. Figure 7.3 illustrates such assumption.

It can be clearly observed that all first order derivatives (slope coefficients of lines tangent to the curve) are always positive at all points of their corresponding definition domains.

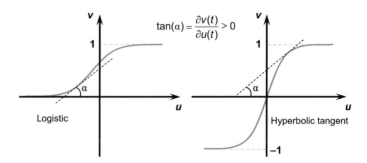

Fig. 7.3   Illustration of first-order partial derivatives

Fig. 7.4 Equilibrium points and their attraction fields

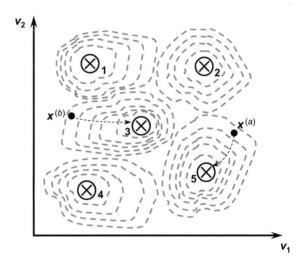

Consequently, there are two essential conditions for the dynamical behavior of the Hopfield Network to be stable, this is:

(i)   The weight matrix $\{W\}$ must be symmetrical.
(ii) The activation function $\{g(.)\}$ must be monotonically increasing.

In summary, by considering that the weight matrix W is symmetrical and that the activation function is monotonically increasing, the demonstration attests that given any data set $x^{(0)}$ of initial conditions, the network converges to a point of stable equilibrium.

Figure 7.4 shows a set of equilibrium points and their corresponding attraction fields, which were obtained by a Hopfield network with two neurons and outputs given by v_1 and v_2.

The five equilibrium points (attractors), represented sequentially in Fig. 7.4, are those to produce the lowest values for the energy function given in (7.5), i.e., they are the states that minimize the value of the network energy function. As an example, as also shown in Fig. 7.4, the network output would converge to equilibrium point number 5 if the network were to be initialized at point $x^{(a)}$ because this point is positioned in the corresponding attraction field. The convergence would occur in equilibrium point number 3 if the network were to be initialized at point $x^{(b)}$.

As the Hopfield network is deterministic, for any initial position that is within an attraction region of an equilibrium point, the network always tends to converge (asymptotically) to such point. The exact number of equilibrium points and their locations are determined by the free parameters $\{W$ and $i^b\}$ of the Hopfield network (Eq. 7.5), being the geometric shapes of the corresponding attraction fields also related to characteristics of the energy function considered.

7.4   Associative Memories

One of the most widespread applications of Hopfield networks are binary associative memories, also called content-addressable memories. The purpose of an associative memory is to recover a pattern, which was previously stored in its structure, from a partial sample (incomplete) or a noisy sample (distorted) of the original pattern (Hopfield 1982). Figure 7.5 illustrates an original version for a particular pattern and two samples with incorrect versions.

In this example, given an incomplete sample (Fig. 7.5b) or a noisy sample (Fig. 7.5c), the objective of the associative memory is to recover the correct original pattern (Fig. 7.5a), which was stored in its structure previously.

Based on the problem exposed above, the Hopfield network can also be used as an associative memory. Therefore, as in all other applications related to the use of Hopfield networks, the challenge is to set up the free parameters of the network in an appropriate way, so that the corresponding energy function gets minimized. In this way, two classical methods are frequently used, namely, the outer product method and the pseudoinverse matrix method.

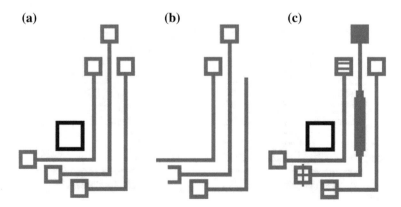

Fig. 7.5 Versions of incorrect patterns in associative memories. **a** Original pattern. **b** Incomplete sample. **c** Noisy sample

7.4.1   Outer Product Method

The simplest way to determine the weight matrix $\boldsymbol{W}$ and the bias vector $\boldsymbol{i}^b$, in order to parametrize associative memories, is the one proposed by Hopfield itself, which was inspired by the application of Hebb's learning method.

According to Hopfield (1982), given a quantity p of patterns $\{z\}$ to be stored in the memory, with n elements each, the free parameters of the Hopfield network are defined by:

$$\boldsymbol{W} = \frac{1}{n}\sum_{k=1}^{p} z^{(k)} \cdot \left(z^{(k)}\right)^{T} \tag{7.10}$$

$$\boldsymbol{i}^b = \boldsymbol{0}, \tag{7.11}$$

where a sign function, as given by expression (1.5), is used as an activation function.

In the case of associative memories, the diagonal of the weight matrix must have null values (indicating the absence of self-feedback). Therefore, by rewriting expression (7.10), it is obtained:

$$\boldsymbol{W} = \frac{1}{n}\sum_{k=1}^{p}\underbrace{z^{(k)} \cdot \left(z^{(k)}\right)^{T}}_{(i)} - \underbrace{\frac{p}{n} \cdot \boldsymbol{I}}_{(ii)}, \tag{7.12}$$

where $\boldsymbol{I} \in \mathfrak{R}^{n \times n}$ is the identity matrix.

It is sure that parcel (i) in expression (7.12) performs the outer product between the elements of each pattern being stored while parcel (ii) simply neutralizes the elements on the main diagonal ($W_{kk} = 0$).

As an example of the assembly of the $\boldsymbol{W}$ matrix, consider two patterns to be stored in an associative memory, as shown in the diagrams of Fig. 7.6.

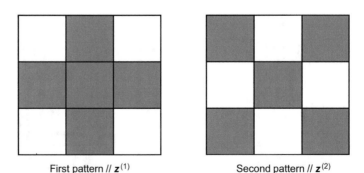

First pattern // $\boldsymbol{z}^{(1)}$ Second pattern // $\boldsymbol{z}^{(2)}$

Fig. 7.6 Examples of patterns to be stored in the memory

In the to-be-stored images, defined by a 3×3 dimension grid, dark grid cells (pixels) are represented by the value $+1$, while white grid cells are represented by the value -1. Such assignments are directly assumed because they correspond to the output values of the sign activation function.

In this condition, vectors $z^{(1)}$ and $z^{(2)}$ are made of the concatenation of lines comprising each corresponding grid, that is:

$$z^{(1)} = \underbrace{\begin{bmatrix} -1 & +1 & -1 \end{bmatrix}}_{\text{1st line}} \underbrace{\begin{matrix} +1 & +1 & +1 \end{matrix}}_{\text{2nd line}} \underbrace{\begin{matrix} -1 & +1 & -1 \end{matrix}}_{\text{3rd line}} \end{bmatrix}^T$$

$$z^{(2)} = \underbrace{\begin{bmatrix} +1 & -1 & +1 \end{bmatrix}}_{\text{1st line}} \underbrace{\begin{matrix} -1 & +1 & -1 \end{matrix}}_{\text{2nd line}} \underbrace{\begin{matrix} +1 & -1 & +1 \end{matrix}}_{\text{3rd line}} \end{bmatrix}^T$$

By using expression (7.12), with $p = 2$ and $n = 9$, the desired weight matrix of the Hopfield network is obtained, so:

$$W = \begin{bmatrix}
0 & -0.22 & 0.22 & -0.22 & 0 & -0.22 & 0.22 & -0.22 & 0.22 \\
-0.22 & 0 & -0.22 & 0.22 & 0 & 0.22 & -0.22 & 0.22 & -0.22 \\
0.22 & -0.22 & 0 & -0.22 & 0 & -0.22 & 0.22 & -0.22 & 0.22 \\
-0.22 & 0.22 & -0.22 & 0 & 0 & 0.22 & -0.22 & 0.22 & -0.22 \\
0 & 0 & 0 & 0 & 0 & 0 & 0 & 0 & 0 \\
-0.22 & 0.22 & -0.22 & 0.22 & 0 & 0 & -0.22 & 0.22 & -0.22 \\
0.22 & -0.22 & 0.22 & -0.22 & 0 & -0.22 & 0 & -0.22 & 0.22 \\
-0.22 & 0.22 & -0.22 & 0.22 & 0 & 0.22 & -0.22 & 0 & -0.22 \\
0.22 & -0.22 & 0.22 & -0.22 & 0 & -0.22 & 0.22 & -0.22 & 0
\end{bmatrix}$$

Therefore, when considering an incomplete or noisy version of the $z^{(1)}$ or $z^{(2)}$ patterns, which will be presented to the network inputs, the network will converge to one of the corresponding original versions stored in its structure.

7.4.2 Pseudoinverse Matrix Method

The assembly of the weight matrix W by using the pseudoinverse matrix method is based on minimizing the error between a partial (or noisy) version and its corresponding original version. Therefore, consider the following matrices:

$$Z = \begin{bmatrix} z^{(1)} & z^{(2)} & z^{(3)} & \cdots & z^{(p)} \end{bmatrix} \tag{7.13}$$

$$R = \begin{bmatrix} r^{(1)} & r^{(2)} & r^{(3)} & \cdots & r^{(p)} \end{bmatrix}, \tag{7.14}$$

where Z is a matrix that contains all p patterns of dimension n, which are represented by vectors $z^{(.)}$. Matrix R contains noisy versions, namely, $r^{(.)}$ of these p patterns.

The existing error between Z and R can be computed by though the Euclidian norm given by:

$$\text{Error} = \|Z - V\|, \tag{7.15}$$

where V is the matrix that represents the network outputs with respect to matrix R, which was presented to the network inputs. In convergence situations, the following relationship stands for the associative memory:

$$V = W \cdot R \tag{7.16}$$

By substituting expression (7.16) in (7.15), it is obtained:

$$\text{Error} = \|Z - W \cdot R\| \tag{7.17}$$

Hence, values for the elements in weight matrix W that minimize expression (7.17) are desired. Through linear algebra and system identification techniques (Ljung 1999), a minimization of this norm can be obtained by:

$$W = Z \cdot (R^T \cdot R)^{-1} \cdot R^T, \tag{7.18}$$

where the parcel $(R^T \cdot R)^{-1} \cdot R^T$ is the pseudoinverse of matrix R, also known as the Penrose-Moore pseudoinverse.

7.4.3 Storage Capacity of Memories

One of the most frequent questions about the Hopfield networks, when used as an associative memory, concerns the number of patterns that can be inserted into their structures when seeking a proper recovery of a pattern from incomplete or noisy samples.

Based on several computational experiments, Hopfield (1982) describes that the storage capacity $\{C^{\text{Hopf}}\}$ of patterns in associative memories, when seeking a pattern recovery with few errors, is given by:

$$C^{\text{Hopf}} = 0.15 \cdot n, \tag{7.19}$$

where n, as already explained, is the number of elements (dimension) of each pattern to be stored.

Table 7.1 Storage capacity of the Hopfield network

Dimension	C^{Hopf}	C^{Max}	$C^{100\%}$
$n = 20$	3.0	3.3	1.7
$n = 50$	7.5	6.4	3.2
$n = 100$	15.0	10.9	5.5
$n = 500$	75.0	40.2	20.1
$n = 1000$	150.0	72.4	36.2

Through probability analysis, in Amit (1992) and Haykin (2009), more accurate results about the storage capacity of the Hopfield network are elaborated and presented. These studies have shown that the maximum storage capacity $\{C^{\text{Max}}\}$ when considering an almost errorless recovery is given by:

$$C^{\text{Max}} = \frac{n}{2 \cdot \ln(n)} \tag{7.20}$$

For a perfect recovery, with accuracy near 100 %, the storage capacity $\{C^{100\%}\}$ of patterns could be estimated by:

$$C^{100\%} = \frac{n}{4 \cdot \ln(n)} \tag{7.21}$$

For comparative purposes, according to the results of the three previous expressions, Table 7.1 shows the storage capacity of the Hopfield network for various values of n.

By analyzing Table 7.1, it is possible to verify that, during a recovery, if the accuracy required is no longer a vital factor, more patterns can be stored in the associative memory as indicated by the C^{Hopf} values. In contrast, when the required accuracy is too high, the maximum number of patterns to be stored $\{C^{\text{Max}}\}$ substantially decays with the increase of the pattern dimension, reaching minimum values when a recovery of 100 % is expected.

Indeed, one of the main weaknesses of the Hopfield network, operating as an associative memory, is its low storage capacity, which is proportional to the dimension of the patterns to be saved.

On the other hand, when the amount of storage recommended in (7.19) and (7.20) is neglected, i.e., an excessive number of patterns are inserted into the network structure, then there is the onset of spurious states. Although they are stable equilibrium points, such states do not correspond to any of the previously stored patterns. In this case, the matrix W has degenerate eigenvalues. A detailed theoretical investigation on this topic, which involves the convergence of the Hopfield network regarding the properties of the matrix W and its subspaces, is examined in Aiyer et al. (1990).

Several other investigations have also been conducted with the purpose of increasing storage capacity in associative memories when using the Hopfield networks, such as those proposed in García and Moreno (2004), Lee (2006), Li et al. (1989) and Muezzinoglu et al. (2005).

7.5 Design Aspects of the Hopfield Network

Unlike the neural architectures presented in previous chapters, the free parameters of the Hopfield network, which are defined by the weight matrix W and the bias vector i^b, are obtained explicitly for most applications. This fact implies in the dismissal of training algorithms.

In fact, the majority of the problems concerning the Hopfield network is derived from the design of the energy function, as defined in expression (7.5), which represents its dynamical behavior.

The more knowledge one has about the study area regarding a particular problem, the more subsidies are available to design its corresponding energy functions.

In summary, a Hopfield network project can be developed with attention to the following aspects:

(i)   The energy function of the problem must be written as in (7.5).

(ii)   For problems concerning system optimization, the energy function is usually formulated through the objective function (cost function or performance function).

(iii)   For constrained optimization problems, both the objective function and structural constraints must be written as expression (7.5). An example of this analysis will be presented in Chap. 20.

(iv)   The weight matrix W may also be composed of algebraic functions. However, these functions must occupy symmetrical positions in W.

(v)   Regardless of the problem being mapped by the Hopfield network, it should always be observed the symmetry in the weight matrix of the network.

(vi)   The activation function of the Hopfield network should always be monotonically increasing (considering the requirements for ensuring stability).

(vii)   In several practical applications, elements of the bias vector i^b can assume null values.

(viii) The range of the domain of each variable being mapped must fall within the limits of the activation function considered.

(ix)   The result of the Hopfield network that corresponds to a solution to the problem being analyzed is represented by one of its equilibrium points.

(x)   In case of associative memory problems, attention should be paid to their storage capacities with respect to the accuracy required for pattern recovery.

The instructions (pseudocode) describing the discrete Hopfield network operation, represented by expressions (7.3) and (7.4), are provided next.

BEGIN {HOPFIELD ALGORITHM – DISCRETE TIME OPERATION}
>   <1> Specify the weight matrix W and the bias vector i^b;
>   <2> Present the initial input vector $\{x^{(0)}\}$;
>   <3> $v^{\text{current}} \leftarrow x^{(0)}$;
>   <4> Repeat the instructions:
>   <4.1> $v^{\text{previous}} \leftarrow v^{\text{current}}$;
>   <4.2> $u \leftarrow W \cdot v^{\text{previous}} + i^b$;   {according to (7.3)}
>   <4.3> $v^{\text{current}} \leftarrow g(u)$;   {according to (7.4)}
>   Until: $v^{\text{current}} \cong v^{\text{previous}}$
>   <5> $v^{\text{final}} \leftarrow v^{\text{current}}$ { v^{final} represents an equilibrium point}

END {HOPFIELD ALGORITHM – DISCRETE TIME OPERATION}

The activation function used in the algorithm depends on the problem being mapped by the Hopfield network, and, in the case of problems involving associative memories, the signal function is the most used one.

In summary, from an initial state $x^{(0)}$, the network tends to converge to one of its stable equilibrium points, where the network progress remains static ($v^{\text{current}} \cong v^{\text{previous}}$).

7.6   Hardware Implementation Aspects

The relatively easy implementation in analog hardware is also one of the main reasons for the popularity of the Hopfield network. Therefore, in Hopfield (1984), besides the proposition of the continuous-time network, it has been demonstrated that the network can be represented by very basic electronic components, such as operational amplifiers, resistors, and capacitors. In this situation, neurons are modeled by operational amplifiers, whose input/output relationship plays the role of the logistic activation function. Figure 7.7 shows the schematic of a hardware implementation of the Hopfield network.

The time constant that defines the progress of each jth Operational Amplifier $\{OA^{(j)}\}$ is specified by the set up of the following internal parameters:

- $R_j^{\text{in}} \rightarrow$ Input Resistor of the jth operational amplifier.
- $C_j^{\text{in}} \rightarrow$ Input Capacitor of the jth operational amplifier.

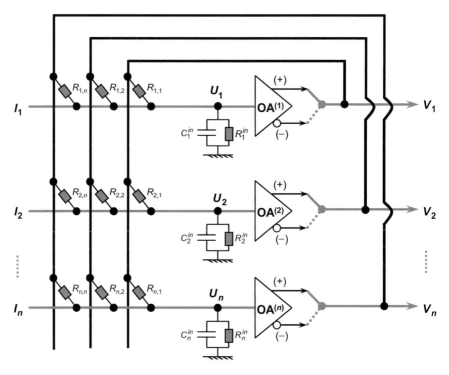

Fig. 7.7 Analog hardware diagram of the Hopfield network

The U_j parameter in Fig. 7.7 indicates the voltage at the input of an operational amplifier and V_j is its output voltage. The network weights W_{ji}, which represent a connection between two neurons, are defined by a conductance connecting the output of the jth amplifier to the input of the ith amplifier. This connection is made using a resistor, whose value is given by:

$$R_{ji} = \frac{1}{|W_{ji}|} \qquad (7.22)$$

If the synapse (connection) is excitatory ($W_{ji} \geq 0$), the resistor is directly connected to the conventional (+) output of the jth operational amplifier; otherwise, if the synapse is inhibitory ($W_{ji} < 0$), the resistor is connected to the inverter (−) output of the corresponding amplifier. Therefore, signals from the elements of the matrix W define the connection points between all operational amplifiers (neurons). The input current of a neuron is formed by the sum of currents flowing through the set of resistors connecting its inputs to other neurons outputs.

As also shown in Fig. 7.7, the equivalent circuit that represents the Hopfield network includes an input current I_j, externally provided for each neuron, which corresponds to the value of the neuron threshold or bias, i.e., $I_j = i_j^b$.

Finally, the equation that describes the progression of the circuit over time can be obtained by applying Ohm's law and Kirchoff's laws (Boylestad and Nashelsky 2012). Assuming that all operational amplifiers have the same internal settings $\{R = R_j^{in}$ and $C = C_j^{in}\}$, it is obtained:

$$C \cdot \frac{dU_j(t)}{dt} = -\eta \cdot U_j(t) + \sum_{j=1}^{n} W_{ji} \cdot V_j(t) + I_j, \text{ with } j = 1, \ldots, n, \quad (7.23)$$

where $\eta = \frac{1}{R} + \sum_{i=1}^{n} \frac{1}{R_{ji}}$.

By normalizing (7.23) with respect to the capacitance value C, one can conclude that the expression is similar to the expression defined in (7.1). Therefore, the electronic diagram shown in Fig. 7.7 corresponds to the equivalent circuit, in analog hardware, of the Hopfield network. Consequently, any practical problem that is mapped using computer simulations can also be converted to an equivalent circuit.

7.7 Exercises

1. What are the essential conditions that must be imposed to the equations that define the dynamics of the Hopfield network so that the network could produce stable solutions?
2. Discuss the importance of the parameter i^b for the convergence process toward stable equilibrium points of the Hopfield network.
3. Explain whether the Gaussian activation function could be used in the neurons of the Hopfield network.
4. Show, in details, what sequence of steps is required to design Hopfield networks that are dynamically stable.
5. Consider a problem regarding associative memories, discuss if the use of the Hopfield network could produce stable equilibrium points that are not a valid solution to the problem.
6. Discuss whether there is any objection to be held to the parameters of the Hopfield network, in discrete time, when the network is initialized with null values $\{x^{(0)} = 0\}$.
7. A team of engineers and scientists have designed a Hopfield network to solve a particular application. However, all elements of the weight matrix had the same values and were different from zero. Explain whether this formulation could be incorrect.
8. Discuss how the activation function hyperbolic tangent should be adjusted, in order to be used in the hardware mapping of associative memories (continuous Hopfield network).

9. Consider an application of associative memories using the Hopfield network, in which each pattern to be stored consists of n dimensions, determine the maximum number of nonzero synaptic weights that the weight matrix could contain.

10. Consider a combinatorial optimization problem that uses a Hopfield network as a solution. In several simulations performed, it was noted that there are only two stable equilibrium points. Discuss on the following items:

   (a) What are the parameters that determine the convergence to one of these equilibrium points?

   (b) Which criterion should be used for choosing which one of these equilibrium points correspond to the final solution of the problem.

7.8   Practical Work

An image transmission system (coded by 45 bits) sends images via a communication link. When arriving at the reception system, the information is decoded in order to recover, accurately, the image previously sent. The four images (information) transmitted are represented in Fig. 7.8.

During the transmission over the communication link, the images are corrupted by noise, and when decoded by the reception system, become incomplete or distorted representations of the image.

To solve this problem, an associative memory is implemented through a Hopfield network, which consists of 45 neurons, to store and recover the images (patterns) defined in Fig. 7.8. Therefore, consider the following conventions:

- White pixel is encoded with value -1.
- Dark pixel is encoded with value $+1$.
- About 20 % of the pixels are randomly corrupted during transmission, i.e., some values -1 becomes $+1$ and vice versa.
- The weight matrix W is defined by the outer product method.

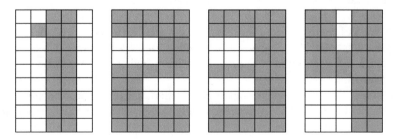

Fig. 7.8 Images sent by the transmitter

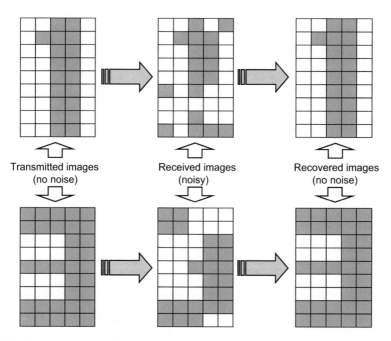

Fig. 7.9 Illustrative scheme of the transmitted, received and recovered images

- The activation function used in all neurons is the signal function (1.5), or the hyperbolic tangent (1.9) function, with the slope parameter $\{\beta\}$ equal to 100.

After the computational implementation of the Hopfield network is finished, do the following exercises.

1. Simulate 12 transmission situations (3 for each pattern) as the illustrative scheme shown in Fig. 7.9.
2. Show the distorted image and the recovered clean image in every situation.
3. Explain what happens when the noise level is increased in an extreme way.

Chapter 8
Self-Organizing Kohonen Networks

8.1 Introduction

However, in some particular applications, only a set of input samples is available, not being available their corresponding desired outputs. On the other hand, these samples have relevant information about the behavior of the system from where they were extracted.

Most networks used in problems with these characteristics organize themselves through competitive training methods, which have the ability to detect similarities, regularities, and correlations between the samples of the input set, grouping them into categories (clusters). Each one of these clusters has particular features related to the conditions governing the operation of the process. Hence, these cluster recognitions are important for understanding the relationships between samples, allowing the identification of the role played by the samples based on attributes of other samples that are part of the same group.

One of the most widespread structures of artificial neural networks regarding the context of self-organization are the self-organizing maps proposed by the Finnish Teuvo Kohonen, which is also known as Kohonen self-organizing networks, or simply Kohonen networks, whose original work was formulated in Kohonen (1982, 1984).

Similarly to the Hopfield network, researches developed by Kohonen attracted attention from the scientific community, at that time and therefore, contributed to a growing revival of the study area of artificial neural networks, which had its definitive revival unfolded with the proposition of the backpropagation algorithm by Rumelhart et al. (1986).

Kohonen networks are inspired by the cerebral cortex, where the activation of a particular region corresponds to the response to a specific sensory stimulus, such as motor, visual, or aural stimulation.

© Springer International Publishing Switzerland 2017
I.N. da Silva et al., *Artificial Neural Networks*,
DOI 10.1007/978-3-319-43162-8_8

Although the Kohonen network is used for various applications in different knowledge areas, the best-disseminated investigations regard problems concerning pattern classification and data grouping (clustering).

8.2   Competitive Learning Process

To describe the steps in the competitive learning process used in the Kohonen network, consider a neural structure consisting of only one neural layer, as illustrated in Fig. 8.1.

The lateral connections between neurons in Fig. 8.1 represent the fact that one neuron can influence the output produced by another neuron. The meaning of these lateral connections will be detailed in the next section.

For simplicity of notation, since the neural structure illustrated in Fig. 8.1 is comprised of a single layer, the following conventions for the weight vectors are assumed:

$$
\begin{aligned}
\mathbf{w}^{(1)} &= \begin{bmatrix} W_{1,1} & W_{1,2} & \dots & W_{1,n} \end{bmatrix}^T && \{\text{Weight vector of the 1st neuron}\} \\
\mathbf{w}^{(2)} &= \begin{bmatrix} W_{2,1} & W_{2,2} & \dots & W_{2,n} \end{bmatrix}^T && \{\text{Weight vector of the 2st neuron}\} \\
\mathbf{w}^{(3)} &= \begin{bmatrix} W_{3,1} & W_{3,2} & \dots & W_{3,n} \end{bmatrix}^T && \{\text{Weight vector of the 3st neuron}\} \\
&(\dots) \\
\mathbf{w}^{(n_1)} &= \begin{bmatrix} W_{n_1,1} & W_{n_1,2} & \dots & W_{n_1,n} \end{bmatrix}^T && \{\text{Weight vector of the last neuron}\}
\end{aligned}
$$

The fundamental principle of the competitive learning process is the competition between neurons aiming a winner neuron, since, as already stated, the process is unsupervised (there are no desired outputs). The prize for those who win the competition between neurons is the weight adjustment, which is proportional to the input values of the pattern presented so to improve their state for the next contest (next sample to be presented).

Fig. 8.1 Basic neural structure of a competitive network

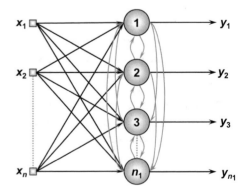

In this circumstance, if all lateral connections of the winner neuron are null (absence of lateral connections), it implies that only its weights will be adjusted, that is, it is assumed the "winner takes it all" strategy. Otherwise, for the situation where the lateral connections from the winner neuron to its neighbors have nonzero values, then a proportional adjustment shall also be made in the weight vector of these adjacent neurons.

Considering this context, it is required to establish a rule that defines which neuron will be the winner neuron. One of the most used methods consists in determining the proximity level between each weight vector of a neuron and the input vector, which contains elements of the kth $\{x^{(k)}\}$ sample presented to the inputs of the network. A typically used proximity metric is the distance between these two parameters:

$$\text{dist}_j^{(k)} = \sqrt{\sum_{i=1}^{n} \left(x_i^{(k)} - w_i^{(j)} \right)^2}, \quad \text{with } j = 1, \ldots, n_1, \tag{8.1}$$

where $\text{dist}_j^{(k)}$ quantifies the Euclidian distance (Euclidian norm) between the input vector representing the kth sample $\{x^{(k)}\}$ and the weight vector of the jth neuron $\{w^{(j)}\}$.

So, the neuron j that obtains the shortest distance is then declared the winner of the competition with respect to sample k. As a reward for victory, the weight vector of the winning neuron is adjusted so that it can approximate itself to the sample. In this way, the following update method is used:

$$w_{\text{current}}^{(v)} = w_{\text{previous}}^{(v)} + \eta \cdot \left(x^{(k)} - w_{\text{previous}}^{(v)} \right) \tag{8.2}$$

In algorithmic notation, expression (8.2) is equivalent to the following:

$$w^{(v)} \leftarrow w^{(v)} + \eta \cdot \left(x^{(k)} - w^{(v)} \right), \tag{8.3}$$

where $w^{(v)}$ represents the weight vector of the winner neuron and the parameter η defines the learning rate.

It is desirable to normalize (in a unitary way) all weight vectors, as well as all sample vectors to obtain a better efficiency of the learning process. The unitary normalization is done by simply dividing each vector by its norm.

Figure 8.2 illustrates the rule for adjusting the weights of the winner neuron in the competitive process. Three neurons (represented by $w^{(1)}$, $w^{(2)}$, and $w^{(3)}$) dispute the tournament with respect to the sample $x^{(k)}$, which is composed of two inputs x_1 and x_2.

By examining this figure and considering that all vectors have already been normalized, it is observed that the vector $w^{(2)}$ is the winner of the competition against $w^{(1)}$ and $w^{(3)}$ since it is the closest vector to the pattern $x^{(k)}$. Consequently, as a reward, its weight vector is updated as shown in expression (8.2), in order to

Fig. 8.2 Adjustment process
of the weight vector of the
winner neuron

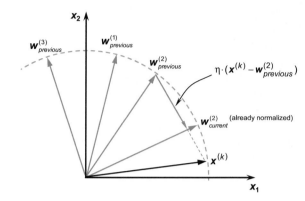

make it even closer to the vector $x^{(k)}$. Undoubtedly, the adjustment consists solely
of rotating the winner neuron toward the vector representing the input sample.

The following instructions describe the algorithmic steps used for the training
phase of the neural structure with competitive learning.

BEGIN {COMPETITIVE ALGORITHM – TRAINING PHASE}

    <1> Obtain the set of training samples $\{x^{(k)}\}$;

    <2> Initially set up the weight vector of each neuron by assigning
             them the first n_1 values of the training samples;

    <3> Normalize the samples and weight vectors;

    <4> Specify the learning rate $\{\eta\}$;

    <5> Start the epoch counter $\{epoch \leftarrow 0\}$;

    <6> Repeat the instructions:

            <6.1> For all training samples $\{x^{(k)}\}$, do:

                  <6.1.1> Calculate the Euclidean distance between $x^{(k)}$
                            and $w^{(j)}$, according to expression (8.1);

                  <6.1.2> Declare as the winner the neuron j with the
                            smallest Euclidean distance;

                  <6.1.3> Adjust the weight vector of the winner neuron
                            according to expression (8.3);

                  <6.1.4> Normalize the weight vector that was adjusted
                            in the previous instruction;

            <6.2> $epoch \leftarrow epoch + 1$;

    Until: there are no significant changes in the weights.

END {COMPETITIVE ALGORITHM – TRAINING PHASE}

Fig. 8.3 Distribution of
weight vectors in a
two-dimensional space with
respect to a set of samples

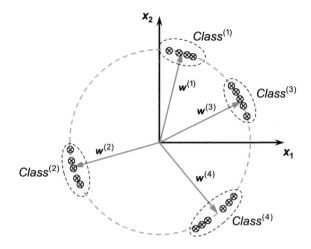

After the convergence of the competitive algorithm, each weight vector is positioned at the center of the clusters, which have similar characteristics. These clusters represent the classes associated with the pattern classification problem, and the number of clusters is automatically linked to the number of neurons used in the neural structure.

It is worth noting that the initial number of neurons used to represent classes with common characteristics is unknown beforehand since the learning process is unsupervised. Hence, it is crucial to obtain further information about the problem being mapped, either from experts or by using statistical methods, which could lead to an initial estimate of the possible number of classes associated with the problem. For the application formulated in Chap. 15, the number of neurons was defined according to the number of possible types of coffee aromas that could be identified by the network.

As an example, consider that four classes are available. Figure 8.3 illustrates how the weight vectors of the network would eventually be distributed after their stabilization. Each of these vectors, which represent the samples of the problem, is comprised of only two input signals $\{x_1$ and $x_2\}$.

It can be noticed in Fig. 8.3, as well as in Fig. 8.2, that all vectors end up being located inside the unit circle, since they are two-dimensional vectors unitarily normalized. It can also be verified that all four weight vectors are positioned at the center of the clusters representing the samples (identified by symbol "$\otimes$"). Consequently, it can also be stated that the neural structure shown in this figure consists of four neurons, since there are only four weight vectors $\{w^{(1)}, w^{(2)}, w^{(3)},$ and $w^{(4)}\}$.

Fig. 8.4 Distribution of
weight vectors in a
three-dimensional space with
respect to a set of samples

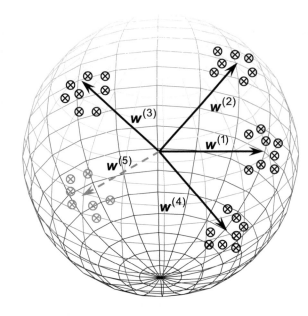

It can be further observed in Fig. 8.3, that if the neural structure was formed by
five neurons, instead of four, then the samples belonging to class (4) would likely be
divided into two classes, since the class is currently gathering two subsets (although
very close) in a single cluster.

For the application shown in Fig. 8.4, each sample available consists of three
input signals, and when they are properly normalized, all sample vectors and weight
vectors are contained in a unitary sphere.

In the case shown in Fig. 8.4, it is clear that those five neurons, represented by
their corresponding weight vectors $\{w^{(1)}, w^{(2)}, w^{(3)}, w^{(4)}, \text{and } w^{(5)}\}$, are positioned in
the middle of five groupings (clusters).

Therefore, through Figs. 8.3 and 8.4, it can be verified that the patterns are
projected onto the space that defines the network weight vectors, whose dimension
is equal to the one in which the patterns are defined. Thus, one of the requirements
for improving the efficiency of the convergence process is that the input variables
should be independent so that its projection occurs on a hypersphere. Otherwise, the
surface stops being spherical, since its axes are no longer orthogonal.

Therefore, after the convergence of the neural structure, when a new sample is
presented to the network for classification purposes, one can simply identify what
neuron is the winner, since it indicates the class to which the sample is closest.

The following instructions describe the algorithmic steps used for the operation
phase of the neural structure with competitive learning.

BEGIN {COMPETITIVE ALGORITHM – OPERATION PHASE}

<1> Present the sample {**x**} to be classified;

<2> Normalize the vector {**x**} that corresponds to the sample;

<3> Adopt the weight vectors {**w** $^{(j)}$} adjusted during the training phase;

<4> Execute the following instructions:

    <4.1> Calculate the Euclidian distances between **x** and **w**$^{(j)}$ according to expression (8.1);

    <4.2> Declare as the winner the neuron *j* with the smallest Euclidean distance;

    <4.3> Associate the sample to the class being represented by the winner neuron;

<5> Provide the class with which the sample was associated.

END {COMPETITIVE ALGORITHM – OPERATION PHASE}

Similarly to the neural structures presented in previous chapters, there is no change in the values of the weight vectors during the operation phase of the competitive network, since those adjustments are only made during the training phase.

8.3   Kohonen Self-Organizing Maps (SOM)

Kohonen self-organizing maps, also known as SOM (Self-Organizing Maps), are considered an architecture of artificial neural networks with (mesh) reticulated structure and competitive learning (Kohonen 1984).

Self-organizing maps are, in its essence, competitive neural structures similar to that shown in Fig. 8.1. Additionally, the lateral connections that represent the influence of the output of the winner neuron to other neurons of the network are provided by topological neighborhood maps. In this situation, the strength of these lateral connections is greater (excitatory) when the neighborhood neurons are closer to the winner neuron, whereas the strength gets smaller as the distance between them gets bigger.

These topological maps express how neurons are spatially organized, with respect to the behavior of their neighbors, and, usually are of one dimension (array) or two dimensions (grid).

Figure 8.5 shows a one-dimensional self-organizing map. As shown in Fig. 8.1, the input vector {**x**}, corresponding to the input samples, is presented to all the neurons in the map. Therefore, here, all the neurons are spatially arranged in a single line.

Fig. 8.5 One-dimensional
topological map

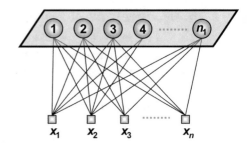

An illustration of a two-dimensional map comprising 16 neurons is shown in Fig. 8.6. Note that in this circumstance the neurons are spatially arranged in rows and columns.

Figure 8.7 shows a matrix representation of the topological map of Fig. 8.6. Although the number of rows and columns could be uneven, for this example, the map is composed of four rows and four columns of neurons.

After the spatial arrangement is defined for constructing the topological map, it is enough to specify the interneuron neighborhood criterion to indicate how neurons cooperate with their neighbors.

One of the most used neighborhood criteria consists of specifying a radius of coverage R, which is used by the neurons of the network to define their corresponding neighbors.

Therefore, for a particular neuron j, its neighbors are the neurons positioned at a maximum distance smaller or equal than R. As an example, consider neuron number 6 (Fig. 8.7). This neuron has as its neighbors the neurons number 2, 5, 7, and 10 when considering a neighborhood radius equal to 1 (inner dashed circle). In the case of a radius equal to 1.75 (outer dashed circle), its neighborhood would then consist of neurons 1, 2, 3, 5, 7, 9, 10, and 11.

For illustration purposes, by assuming a neighborhood radius equal to one ($R = 1$) for all neurons, Fig. 8.8 shows a topological map with the interconnections between each neuron and its corresponding neighbors.

Fig. 8.6 Two-dimensional
topological map

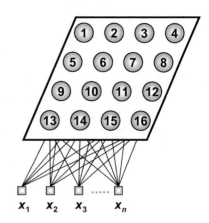

Fig. 8.7 Matrix
representation of the
two-dimensional topological
map

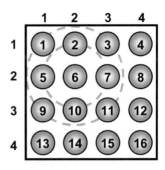

Neighborhood sets are represented by the parameter $\Omega_j^{(R)}$, which is defined for all neurons in the topological map. This parameter indicates which are the neurons in the neighborhood of the neuron j, whose neighborhood radius is R. In Fig. 8.8, values for some neighborhood sets are presented when a radius equal to 1 is assumed ($R = 1$). If a radius equal to 1.75 is assumed, then, the set becomes:

$$\Omega_1^{(1.75)} = \{2, 5, 6\} \qquad ; \quad \Omega_2^{(1.75)} = \{1, 3, 5, 6, 7\}$$
$$\Omega_3^{(1.75)} = \{2, 4, 6, 7, 8\} \qquad ; \quad \Omega_4^{(1.75)} = \{3, 7, 8\}$$
$$\Omega_5^{(1.75)} = \{1, 2, 6, 9, 10\} \qquad ; \quad \Omega_6^{(1.75)} = \{1, 2, 3, 5, 7, 9, 10, 11\}$$
$$(\ldots)$$
$$\Omega_{15}^{(1.75)} = \{10, 11, 12, 14, 16\} \quad ; \quad \Omega_{16}^{(1.75)} = \{11, 12, 15\}$$

Hence, consider that the neuron j won the competition when a sample was presented, then its weight vector, as well as the weight vectors of its neighbors will be adjusted. However, the neurons that are in the neighborhood of the winner neuron are adjusted with lower rates than those rates used for adjusting the weights of the winner neuron. A reasonable approach, when using a unitary radius, has been to assume an adjustment value for the weight vector of neurons neighboring the winner equal to half of the learning rate. In this condition, expression given by (8.3) shall be represented by two rules:

$$\begin{cases} \text{Rule 1}: \ w^{(v)} \leftarrow w^{(v)} + \eta \cdot \left(x^{(k)} - w^{(v)}\right), \text{ for the winner neuron} \\ \text{Rule 2}: \ w^{(\Omega)} \leftarrow w^{(\Omega)} + \frac{\eta}{2} \cdot \left(x^{(k)} - w^{(\Omega)}\right), \text{ for the neighboring neurons} \end{cases}, \qquad (8.4)$$

Fig. 8.8 Neighborhood
interconnections when
considering an unitary radius
for each neuron in the
self-organizing map

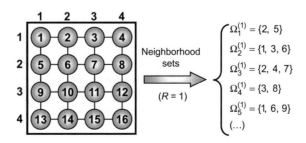

where the first rule is applied to the winner neuron, whereas the second rule is applied to all neurons that are neighbors to the winner. These neurons are specified by the corresponding neighborhood set.

As an example, consider the self-organizing map in Fig. 8.8, if neuron number 10 was the winner, its weight vector would be adjusted by the first rule in expression (8.4) and the weight vectors of its neighboring neurons {6, 9, 11, and 14} would be adjusted by the second rule.

For neighboring regions that are greater than unity, the learning rate could be weighted by a Gaussian function, so that, the greater the distance between the winner neuron and its neighbors, the smaller the adjustments in the weight vectors. In such case, expression (8.4) becomes:

$$\begin{cases} \text{Rule 1}: \ w^{(v)} \leftarrow w^{(v)} + \eta \cdot \left(x^{(k)} - w^{(v)}\right), \text{ for the winner neuron} \\ \text{Rule 2}: \ w^{(\Omega)} \leftarrow w^{(\Omega)} + \eta \cdot \alpha^{(\Omega)} \cdot \left(x^{(k)} - w^{(\Omega)}\right), \text{ for its neighbors} \end{cases} \quad (8.5)$$

where $\alpha^{(\Omega)}$ is a neighborhood operator given by:

$$\alpha^{(\Omega)} = e^{\frac{-\left\| w^{(v)} - w^{(\Omega)} \right\|^2}{2\sigma^2}} \quad (8.6)$$

Therefore, the use of a Gaussian operator provides a decay rate that is strong for those neurons that require a weight adjustment, since they are in the neighborhood set, but are farthest to the winner. Figure 8.9 shows an illustration of this Gaussian decay.

As an example, consider the topological arrangement in Fig. 8.7, with a radius equal to 1.75 (outer dashed circle), the neighborhood set for neuron 6 would be given by: $\Omega_6^{(1.75)} = \{1, 2, 3, 5, 7, 9, 10, 11\}$. In such circumstance, the adjustments of the weight vectors of neurons 2, 5, 7, and 10 would be bigger than adjustments made in the weight vectors of neurons 1, 3, 9, and 11, since they are farther away from the winner.

Fig. 8.9 Illustration of a Gaussian decay for the weight adjustments of neighboring neurons

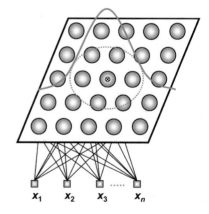

Additionally, from the studies reported in Ritter et al. (1992), another possible strategy aiming an efficiency increase in the adjustment process of the weight vectors consists in gradually decreasing the learning rate during the training process.

Computational instructions of the training algorithm of Kohonen self-organizing maps are presented next.

BEGIN {KOHONEN ALGORITHM – TRAINING PHASE}

<1> Define the network topological map;

<2> Assemble the neighborhood sets $\{\Omega_j^{(R)}\}$;

<3> Obtain the set of training samples $\{\boldsymbol{x}^{(k)}\}$;

<4> Initially set up the weight vector $\{\boldsymbol{w}^{(j)}\}$ of each neuron with the values of the n_1 first training samples;

<5> Normalize the sample vectors and the weight vectors;

<6> Specify the learning rate $\{\eta\}$;

<7> Start the epoch counter $\{epoch \leftarrow 0\}$;

<8> Repeat the instructions:

<8.1> For all training samples $\{\boldsymbol{x}^{(k)}\}$, do:

<8.1.1> Calculate the Euclidian distance between $\boldsymbol{x}^{(k)}$ and $\boldsymbol{w}^{(j)}$, according to expression (8.1);

<8.1.2> Declare as the winner the neuron j that has the smallest Euclidean distance:

$$winner = \arg\min_{j}\{\|x^{(k)} - w^{(j)}\|\}$$

<8.1.3> Adjust the weight vector of the winner neuron according to the rule number 1 in expression (8.4);

<8.1.4> Adjust the weight vectors of the neighborhood neurons, defined in $\Omega_j^{(R)}$, according to the rule number 2 in expression (8.4) or (8.5);

<8.1.5> Normalize the weight vectors that were adjusted in the previous instruction;

<8.2> $epoch \leftarrow epoch + 1$;

Until: there is no significant change in the weight vectors;

<9> Analyze the map in order to extract characteristics;

<10> Identify regions that allow the definition of classes.

END {KOHONEN ALGORITHM – TRAINING PHASE}

Fig. 8.10 Context map for a
two-dimensional topology

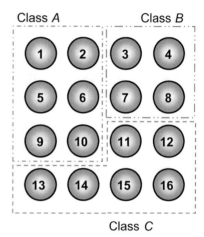

As self-organizing maps do not have the desired outputs (classes) a priori, steps
<9> and <10> of the training phase require the use of statistical tools and
knowledge from experts to identify possible classes of the data. As a consequence,
the topological map can be partitioned into regions that outline the classes, being
such structure called a context map. For example, Fig. 8.10 shows a possible
contextual map for the two-dimensional topology illustrated in Fig. 8.7. This
contextual map was built from analysis made after the training process.

For the example shown in this figure, it is observed that the context map is
characterized by three classes of the problem, which are composed of the following
set of neurons:

$$\text{Class } A \Rightarrow \{1, 2, 5, 6, 9, 10\}$$
$$\text{Class } B \Rightarrow \{3, 4, 7, 8\}$$
$$\text{Class } C \Rightarrow \{11, 12, 13, 14, 15, 16\}$$

Therefore, when presenting a new sample for classification, it is sufficient to find
the winner neuron of the competition. After defining the winner, the contextual map
is used to point out which class the neuron is representing. Finally, the sample is
then classified as belonging to the class corresponding to the winner neuron.

Therefore, it can be stated that contextual maps could also be used as a preprocessing tool for other pattern classification techniques or system identification
techniques.

The following instructions exhibit the algorithmic steps for the operation phase
of the Kohonen maps.

BEGIN{ KOHONEN ALGORITHM – OPERATION PHASE }

<1> Present the sample {x} to be classified and normalize it;

<2> Adopt the weight vectors {$w^{(j)}$} already adjusted during the training phase;

<3> Execute the following instructions:

<3.1> Calculate the Euclidean distance between x and $w^{(j)}$, according to expression (8.1);

<3.2> Declare as the winner the neuron j with the smallest Euclidean distance;

<3.3> Find the winner neuron in the self-organizing map;

<3.4> Associate the sample to the class identified through the context map;

<4> Provide the class to which the sample was associated.

FIM {ALGORITMO KOHONEN – FASE DE OPERAÇÃO}

In summary, by what has been exposed in this section, three important aspects must be considered in order to configure a Kohonen self-organizing map, this is:

- Definition of the spatial arrangement (grid) of the neurons.
- Definition of the neighborhood set of each neuron.
- Specification of the criteria for the weight adjustment of the winner neuron and neighborhood neurons.

Finally, the entire dynamic that comprises the training process of the Kohonen network can be further systematized by the following steps:

1. Each neuron of the network computes the proximity level of its weight vector to each input pattern.
2. A competition mechanism between neurons is applied in order to choose the winner.
3. By using the topological map, it is constructed a neighborhood set that informs which are the neighbors of the winner neuron.
4. Weights of the winner neuron and weights of the neighborhood neurons are incremented so to improve their proximity level to the sample.
5. After convergence, it is possible to identify regions on the contextual map that correspond to the classes of the problem.

Their structures with a simple configuration, as well as their distinguished training dynamics, makes the Kohonen self-organizing maps a sophisticated tool for applications in problems regarding pattern classification and data grouping (clustering).

As also observed, the competitive learning algorithm has some similarity with the k-means algorithm designed by Steinhaus (1956), wherein the available samples are divided into k groups (represented by their centers), and the allocation criterion is the Euclidean norm between the sample and the centers.

However, by what is described in Sect. 6.2.1, one of the limitations of the k-means method is the necessity of specifying, a priori, the number of groups considered for solving the problem. In contrast, the competitive learning algorithm implemented in self-organizing maps facilitates the identification of the number of classes after the convergence of the network.

8.4   Exercises

1. Explain why it is recommended to normalize the weight and sample vectors when the competitive learning method is applied to a network.
2. Write about the possible difficulties one might face during the convergence process of the competitive network when using the weight and sample vectors without normalizing them.
3. What is the meaning of the learning rate parameter in the training phase of the competitive network? Write about the implications of assuming really small values, as well as, really big values for this parameter.
4. Considering expression (8.2), show, step by step (in a geometric form), how the weight adjustment of a neuron is processed.
5. For the topological map depicted in Fig. 8.7, specify the neighborhood set of all neurons with a neighborhood radius equal to 2 ($R = 2$).
6. What is the role of the lateral connections between neurons in a Kohonen self-organizing map?
7. Explain what are the main differences between the competitive neural network presented in Fig. 8.1 and SOM networks.
8. Regarding the diagrams in Figs. 8.3 and 8.4, write about what could happen in these distributions if the specified number of neurons were smaller than the corresponding number of classes.
9. Explain why the design of an SOM network can be associated with some structural characteristics of the cerebral cortex.
10. By considering a specific sample $x^{(k)}$, demonstrate that the rule for adjusting weight vectors based on the Euclidian norm is obtained from the minimization of the squared error function defined by:

$$\text{Error} = \frac{1}{2} \sum_i^n \left(w_i^{(v)} - x_i^{(k)} \right)^2,$$

where $w^{(v)}$ is the winner neuron and n is the number of input signals that composes the sample.

8.5   Practical Work

In the manufacturing process of tires, it is known that the compound that creates the rubber can exhibit imperfections that could prevent its use. Several samples of these anomalies were collected, and several measurements were made for three quantities $\{x_1, x_2,$ and $x_3\}$ that are part of the manufacturing process of the rubber. However, the team of scientists and engineers do not have technical discernment on how these variables are related.

Therefore, a Kohonen network consisting of 16 neurons (Fig. 8.11) is used to detect any existing similarities and correlations between these variables. The final objective is to group the imperfect samples in classes.

Hence, based on the data given in Appendix E, train a Kohonen self-organizing map according to the spatial structure shown in Fig. 8.11. Use a learning rate of 0.001.

The schematic diagram of the two-dimensional grid, which shows the interneuron neighborhoods when assuming a unity neighborhood radius ($R = 1$), is presented in Fig. 8.12.

By using the results from the network training process, it was verified that samples from 1 to 20, from 21 to 60, and from 61 to 120 have some common

Fig. 8.11 Spatial structure of the self-organizing map used in the practical project

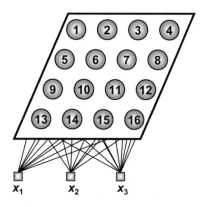

Fig. 8.12 Neighborhood between neurons ($R = 1$)

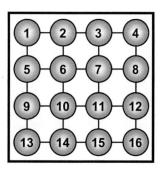

Table 8.1 Samples for class identification by the self-organizing map

Sample	x_1	x_2	x_3	Class
1	0.2471	0.1778	0.2905	
2	0.8240	0.2223	0.7041	
3	0.4960	0.7231	0.5866	
4	0.2923	0.2041	0.2234	
5	0.8118	0.2668	0.7484	
6	0.4837	0.8200	0.4792	
7	0.3248	0.2629	0.2375	
8	0.7209	0.2116	0.7821	
9	0.5259	0.6522	0.5957	
10	0.2075	0.1669	0.1745	
11	0.7830	0.3171	0.7888	
12	0.5393	0.7510	0.5682	

characteristics, and they can be considered belonging to three distinct classes, namely, class A, B, and C respectively.

Consider the following exercises:

1. Identify which set of neurons shown in the grid of Fig. 8.12 corresponds to class A, class B, and class C.
2. For samples shown in Table 8.1, indicate the class to which each of them belongs.

Chapter 9
LVQ and Counter-Propagation Networks

9.1 Introduction

In this chapter, two architectures of artificial neural networks with supervised learning are presented. One of them, the LVQ (Learning Vector Quantization) network has, in its learning process, certain similarities to the Kohonen self-organizing network, and it has usually been applied to pattern classification problems. The training stage of this neural network, which in its standard form is composed of a single neural layer, performs a vector quantization process with respect to the space where input samples are located, in order to adjust the domain regions of each one of the classes. A detailed presentation of the vector quantization process will be developed in the next section of this chapter.

The other artificial neural network analyzed in this chapter is called counter-propagation network. It is constructed of two neural layers. The training process of these layers is a combination of self-organizing learning (used in the first layer) and supervised learning (applied to the second layer). However, the counter-propagation network is classified as a supervised training architecture, since it needs, a priori, a set of input/output pairs to adjust its internal parameters.

Differently from the LVQ network, the counter-propagation network can be used in feature extraction (aiming pattern classification) and in problems related to curve fitting and signal processing, such as those in Chang and Su (2005), Morns and Dlay (2003), Yang et al. (2004), and Lin et al. (1990).

9.2 Vector Quantization Process

Consider a pattern classification problem where samples are divided into n classes. The vector quantization process consists of assigning a unique vector to each one of these classes, namely the quantizing vectors or code vectors. These vectors will

© Springer International Publishing Switzerland 2017
I.N. da Silva et al., *Artificial Neural Networks*,
DOI 10.1007/978-3-319-43162-8_9

Fig. 9.1 Illustration of a
class quantizing vector

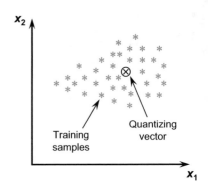

represent the pattern of these groups when the classification of a new sample is issued. Therefore, this new sample is classified as belonging to the class that is closest to the quantizing vector.

For a particular class, the objective of the vector quantization process is to obtain a quantizing vector allocated in a strategic position and that the summation of its distances to all the samples in the class is the smallest. Figure 9.1 depicts the idea that comprises the vector quantization process with respect to a set of samples with two components $\{x_1$ and $x_2\}$.

In Fig. 9.1, it can be observed that the quantizing vector is positioned in a place where the summation of its distance to the elements of the class is as small as possible. In the illustrated situation, it is verified that the quantizing vector is allocated near the place of greatest density of samples. On the other hand, if the samples of a class had a normal distribution then the quantizing vector would tend to be the average of their values.

Now, consider the existence of n classes, then, there will be n quantizing vectors responsible for representing the profile of each class. Figure 9.2 depicts five distinct classes with their corresponding quantizing vectors, which were already calculated.

Hence, each quantizing vector depicted in Fig. 9.2 is representing a specific profile that overlooks each class they are representing.

Fig. 9.2 Illustration of
quantizing vectors

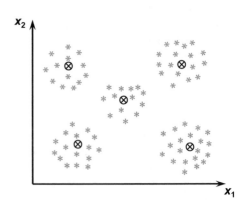

Fig. 9.3 Illustration of the dominant regions of the quantizing vectors

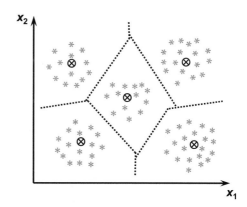

Now, consider that a new sample must be classified among one of the five classes defined in Fig. 9.2. The parameters used for accomplishing this task are the quantizing vectors that represent each class. In this way, the main metric used to classify new samples is the proximity level, where the quantizing vector closest to the sample determines the class to allocate the sample. In fact, the closer the quantizing vector is to the sample, the more they are alike.

Consequently, by using this minimum distance criterion, in order to define the prevailing quantizing vector with respect to a sample, it is possible to outline the boundaries of the quantizing vectors relative to the other quantizing vectors. Given that a midpoint between any two quantizing vectors is a natural proximity limit, it would be enough to find the hyperplane that is perpendicular to the line that connects these vectors, and, as the main condition, impose that the hyperplane passes through the midpoint. As an illustration, Fig. 9.3 shows the prevailing boundaries of the classes associated to each quantizing vector defined in Fig. 9.2.

It can be observed, when assuming the proximity criterion, that the dominant boundaries make it possible to code the entire sample space in subregions, which set the limits between the classes. The complete set of reticles derived from a partitioning, which also allows a dimensionality reduction is known as the Voronoy diagram (Aurenhammer 1991) or the Dirichlet mosaic (Bowyer 1981).

In practical terms, after the allocation of the quantizing vectors, the decision-making of which class does a new sample belong to is done by calculating its distance to all the quantizing vectors. The sample is classified as belonging to the class being represented by the quantizing vector with the shortest distance.

Accordingly, the primary task of the vector quantization process is to determine the spatial position of each quantizing vector within each class of the sample space, and this is also the objective of the LVQ network.

In comparison, vector quantization is very different from clustering. The only purpose of the latter is to find and group samples with common characteristics into classes. In contrast, besides finding groups with similar characteristics, vector quantization consists mainly in converting all the input space, where the samples are defined, into a discrete output space. In this condition, the output turns to be a discrete representation of the input space, which causes a noticeable dimensionality reduction.

9.3   LVQ Networks (Learning Vector Quantization)

The LVQ network was also created by Teuvo Kohonen and is considered a supervised version of self-organizing maps (SOM). Therefore, for training purposes, unlike in the SOM topology, it a set of input/output pairs that represents the process being analyzed is required.

The training process of the LVQ network is also performed in a competitive manner, similarly to those processes used for training SOM networks. The weight vectors of the neurons will represent the corresponding quantizing vectors as illustrated in Sect. 9.2. Therefore, to use this topology, all different classes associated with the problem being analyzed must be known a priori.

Figure 9.4 shows an LVQ network composed of n inputs and n_1 neurons, which represent all the classes in the problem of pattern classification being analyzed.

As shown in Fig. 9.4, in contrast to the Kohonen architecture, the LVQ network has no lateral connections between neurons, and this aspect implies that those neurons that are neighbors to the winner neuron will no longer have their weights adjusted. In fact, the existence of lateral connections is unnecessary, since the training algorithm computes the excitatory and inhibitory stimuli.

Fig. 9.4 Basic structure of an LVQ network

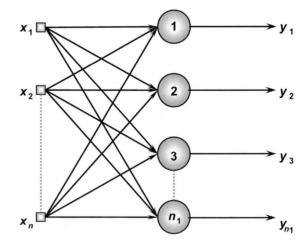

Two training algorithms called LVQ-1 and LVQ-2 are typically used to adjust the weights of the winner neuron. The LVQ-1 adjusts only the weights of the winner neuron while the LVQ-2 algorithm adjusts the weights of both the winner and the vice-winner neurons. Other elaborate procedures for these adjustments, inspired by these algorithms, have also been proposed in the literature, in order to optimize the process in particular applications, such as those presented in Lee and Kim (1995) and Vakil-Baghmisheh and Pavesic (2003).

9.3.1 LVQ-1 Training Algorithm

Through the competitive process involving each one of the available samples, the LVQ-1 training algorithm adjusts only the synaptic weights of the winner neuron.

In this case, it is considered that each input vector $\{x^{(k)}\}$ used during the LVQ training process belongs to only one of the j classes known, since the learning mechanism is supervised.

The inherent algorithmic procedures of the training process are, therefore, identical to those of the Kohonen network, except by the criterion to select which neurons should have their weights adjusted. In the LVQ-1 training algorithm, only the weights of the winner neurons are properly tuned.

The two main steps of the algorithm consist of obtaining a winner neuron as well as specifying the rule for the adjustment of its synaptic weight.

In regards to obtaining the winner, the neuron with the weight vector $\{w^{(j)}\}$ with the biggest proximity level to a particular sample $\{x^{(k)}\}$ is declared victorious. Similarly to the Kohonen network, one of the most used metrics for calculating the proximity level is the Euclidian norm, this is:

$$\text{dist}_j^{(k)} = \sqrt{\sum_{i=1}^{n} \left(x_i^{(k)} - w_i^{(j)} \right)^2}, \quad \text{with } j = 1, \ldots, n_1, \tag{9.1}$$

where $\text{dist}_j^{(k)}$ provides the distance between the input vector which represents the kth sample $\{x^{(k)}\}$ and the weight vector $\{w^{(j)}\}$ of the jth neuron.

After proclaiming the winner, the rule for adjusting the weights is applied. For the situation in which the winner neuron $\{w^{(j)}\}$ is representing the class assigned to the sample $\{x^{(k)}\}$, this is, $x^{(k)} \in C^{(j)}$, the weights are adjusted so as to approximate the winner to the sample. In this circumstance, the reward given to the winner neuron will be a bigger chance to win the competition when that same sample is presented again during the training process. Otherwise, for the condition which the

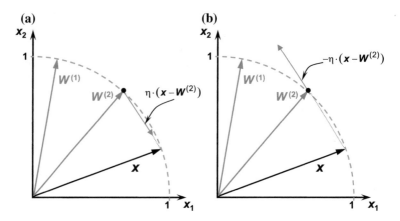

Fig. 9.5 Mechanism for adjusting weights of the LVQ-1 algorithm

winner neuron $\{w^{(j)}\}$ is not representing the class of the sample $\{x^{(k)}\}$, this is, $x^{(k)} \notin C^{(j)}$, then, the weights are adjusted in order to move the winner away from the sample. Consequently, other neurons can now win the competition in the next training epoch. In algorithmic terms, these procedures can be synthesized by the following rules:

$$\text{If } x^{(k)} \in C^{(j)}$$
$$\text{Then: } w^{(j)} \leftarrow w^{(j)} + \eta \cdot (x^{(k)} - w^{(j)}) \tag{9.2}$$
$$\text{Else: } w^{(j)} \leftarrow w^{(j)} - \eta \cdot (x^{(k)} - w^{(j)}),$$

where the parameter η defines the learning rate.

The following figure illustrates the mechanism for adjusting the weights of the proclaimed winner neuron when applying the LVQ-1 training. It is considered that a sample $\{x^{(k)}\}$ is composed of two components $\{x_1$ and $x_2\}$ and belongs to one of the two classes $\{C^{(1)}$ or $C^{(2)}\}$ of the problem.

Hence, suppose that the sample belongs to class type 2, this is, $x^{(k)} \in C^{(2)}$, then the weight vector of neuron 2 $\{w^{(2)}\}$, which represents the corresponding class, is then moved toward the vector $x^{(k)}$, as shown in Fig. 9.5a. Otherwise, in the situation which the sample $x^{(k)} \notin C^{(2)}$, the weight vector $\{w^{(2)}\}$ of the winner neuron must move away from $x^{(k)}$ so that another neuron could win the proximity contest in next iterations.

In summary, it can be stated that if the winner neuron $w^{(2)}$ is representing the class to which the sample $x^{(k)}$ belongs, then, it will be attracted toward $x^{(k)}$; if not, the vector $w^{(2)}$ is repealed, giving a chance to the neuron representing the correct class to win in subsequent steps.

The basic steps of the LVQ-1 algorithm (training phase) are shown next by pseudocode instructions.

BEGIN {LVQ-1 ALGORITHM – TRAINING PHASE}

<1> Obtain a set of training samples $\{x^{(k)}\}$;

<2> Associate to each sample $\{x^{(k)}\}$ its corresponding desired class (output) $\{C^{(j)}\}$, with j varying from one to the total number of classes;

<3> Initialize each weight vector $\{w^{(j)}\}$ with one of the sample values in $\{C^{(j)}\}$;

<4> Normalize the sample vectors $\{x^{(k)}\}$ and weight vectors $\{w^{(j)}\}$;

<5> Specify the learning rate $\{\eta\}$;

<6> Start the epoch counter $\{epoch \leftarrow 0\}$;

<7> Repeat the instructions:

    <7.1> For all training samples $\{x^{(k)}\}$, do:

        <7.1.1> Calculate the Euclidean distances between $x^{(k)}$ and $w^{(j)}$, according to expression (9.1);

        <7.1.2> Declare as the winner the neuron j that has the smallest Euclidean distance:

$$winner = \arg\ \min_{j}\{\|x^{(k)} - w^{(j)}\|\};$$

        <7.1.3> Adjust the weight vector of the winner neuron according to expression (9.2);

        <7.1.4> Normalize the weight vector adjusted in the previous instruction;

    <7.2> $epoch \leftarrow epoch + 1$;

  Until: there is no significant change in the weight vectors.

END {LVQ-1 ALGORITHM – TRAINING PHASE}

After the training phase, the operation phase of the network, when a new sample is presented, consists of defining the winner neuron, whose label is representing the class that will be assigned to the sample. The following algorithm provides the basic instructions for the operation phase.

BEGIN {LVQ-1 ALGORITHM – OPERATION PHASE}

<1> Present the sample $\{x\}$ to be classified and normalize it;

<2> Adopt the weight vectors $\{w^{(j)}\}$ already adjusted during the training phase;

<3> Execute the following instructions:

    <3.1> Calculate the Euclidian distances between x and $w^{(j)}$ according to expression (9.1);

    <3.2> Declare as the winner the neuron j with the smallest Euclidean distance;

    <3.3> Associate the sample to the class being represented by the winner neuron;

<4> Provide the class with which the sample was associated.

END {LVQ-1 ALGORITHM – OPERATION PHASE}

According to what was mentioned above and from the inspection of the instructions in the LVQ-1 algorithm, it can be observed some similarity with the Kohonen algorithm. A major difference is that the classes in which each sample is assigned are known a priori.

9.3.2 LVQ-2 Training Algorithm

In the LVQ-2 training algorithm, the adjustment procedures are applied both to the winner and to the vice-winner neuron.

Consider that the weight vector of the winner neuron is defined by $w^{(j)}$ and the weight vector of the vice-winner is given by $w^{(m)}$. Similarly to the LVQ-1 algorithm, for the situation where the winner neuron is representing the class assigned to the sample $\{x^{(k)}\}$, this is, $x^{(k)} \in C^{(j)}$, the weights are adjusted in order to approximate the winner to the sample. In contrast, and by acting up in advance, the weight vector of the vice-winner is repealed from $x^{(k)}$, since it represents another class, which is different from the one assigned to $x^{(k)}$, this is, $x^{(k)} \notin C^{(m)}$. In algorithmic terms, the following rule is considered:

$$\text{If } (x^{(k)} \in C^{(j)}) \text{ AND } (x^{(k)} \notin C^{(m)})$$
$$\text{Then: } \begin{cases} w^{(j)} \leftarrow w^{(j)} + \eta \cdot (x^{(k)} - w^{(j)}) \\ w^{(m)} \leftarrow w^{(m)} - \eta \cdot (x^{(k)} - w^{(m)}) \end{cases} \qquad (9.3)$$

Alternatively, if the vice-winner neuron is representing the class to which the sample belongs, then, it is enough to change the adjustment rule, which will now approximate the vice-winner, as well as, repeal the winner, so:

$$\text{If } (x^{(k)} \notin C^{(j)}) \text{ AND } (x^{(k)} \in C^{(m)})$$
$$\text{Then: } \begin{cases} w^{(j)} \leftarrow w^{(j)} - \eta \cdot (x^{(k)} - w^{(j)}) \\ w^{(m)} \leftarrow w^{(m)} + \eta \cdot (x^{(k)} - w^{(m)}) \end{cases} \qquad (9.4)$$

Another improvement that can be inserted in the LVQ-2 algorithm, called the LVQ-2.1 training algorithm (Kohonen 1990), consists of applying the above rules only if the sample $x^{(k)}$ is contained inside a membership window λ defined around the borders between $w^{(j)}$ and $w^{(m)}$ that is:

$$\min\left(\frac{d_j}{d_m}, \frac{d_m}{d_j}\right) > \frac{1 - \lambda}{1 + \lambda}, \text{ with } \lambda \in [0.2, 0.3], \qquad (9.5)$$

where d_j and d_m are respectively the distance from $x^{(k)}$ to $w^{(j)}$ and $w^{(m)}$. Therefore, this modification of the LVQ-2 algorithm moves vectors $w^{(j)}$ and $w^{(m)}$ to the vicinity of the boundaries separating the classes, since these boundaries are the ones that minimize the classification error (Bayesian boarders). The enforcement of the condition in (9.5) makes the network much more robust for purposes of pattern classification when its training is finished. Additionally, any possible instabilities in the convergence processes observed during the training phase are also minimized.

The sequence of instructions related to the LVQ-2 algorithm are presented next.

BEGIN {LVQ-2 ALGORITHM – TRAINING PHASE}

<1> Obtain the set of training samples $\{x^{(k)}\}$;

<2> Associate each sample $\{x^{(k)}\}$ with its corresponding desired class (output) $\{C^{(j)}\}$, with j varying from one to the total number of classes;

<3> Initialize each weight vector $\{w^{(j)}\}$ with one of the values from the samples in $\{C^{(j)}\}$;

<4> Normalize the sample $\{x^{(k)}\}$ and weight $\{w^{(j)}\}$ vectors;

<5> Specify the learning rate $\{\eta\}$;

<6> Start the epoch counter $\{epoch \leftarrow 0\}$;

<7> Repeat the instructions:

<7.1> For all training samples $\{x^{(k)}\}$, do:

<7.1.1> Calculate the Euclidean distances between $x^{(k)}$ and $w^{(j)}$, according to expression (9.1);

<7.1.2> Obtain the winner and vice-winner neurons;

<7.1.3> Adjust the weight vector for the winner neuron and for the vice-winner according to rules shown in (9.3) and (9.4);

<7.1.4> Normalize the weight vector adjusted in the previous instruction;

<7.2> $epoch \leftarrow epoch + 1$;

Until: there are no significant changes in the weights.

END {LVQ-2 ALGORITHM – TRAINING PHASE}

The instructions associated with the operation phase of the LVQ-2 algorithm are the same as those used for the LVQ-1.

Furthermore, as verified in the training algorithm of the Kohonen network, the learning rate $\{\eta\}$ in algorithms LVQ-1 and LVQ-2 can also be decreased as the training process advances, avoiding eventual instabilities in the convergence.

In addition to LVQ-1 and LVQ-2 algorithms, there is the proposition of the LVQ-3 algorithm, found in Kohonen (2000), in which additional procedures are formulated to improve the convergence process of quantizing vectors related to the classes of the problem being analyzed.

9.4   Counter-Propagation Networks

Networks with the counterpropagation architecture were designed by Hecht-Nielsen (1987a) and usually consist of two neural layers, whose training processes are considered hybrid, such as the processes used on RBF networks. Figure 9.6 shows a counter-propagation network that consists of n inputs, n_1 neurons in the hidden layer and n_2 neurons in its output layer.

As its name suggest, differently from MLP networks, counter-propagation networks do not make any backpropagation of the error in order to adjust the weights of the neurons in the hidden layer. According to the structural arrangement shown in Fig. 9.6, the first layer of the counter-propagation network is typically a Kohonen structure, and its training proceeds in a similar way to the competitive learning algorithm presented in Sect. 8.2.

As to the second neural layer (output layer), its neurons compose a Grossberg arrangement, also called as the Outstar structure (Grossberg 1974). These neurons have a ramp activation function and absence of bias. The training of this layer is processed in a supervised manner, seemingly similar to Hebb's rule as explained in the following subsection.

Therefore, the training of the counterpropagation network is done in two stages. First, the hidden layer is adjusted in order to identify data clusters. Then, the output layer (outstar) is trained in order to associate the responses of the network with its corresponding desired outputs. To this end, the inputs of the output layer will be the values produced by the hidden layer (Kohonen), where just one winner neuron exists with respect to an input stimulus.

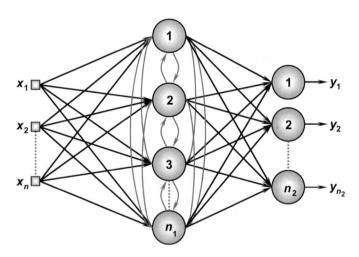

Fig. 9.6 Basic architecture of a counter-propagation network

9.4.1 Aspects of the Outstar Layer

The training process of the outstar layer of a counter-propagation network must be done only after the learning process of the Kohonen layer is finished.

The procedures for adjusting the weights of the neurons in the outstar layer are fairly compact, and only the weights of those neurons with non-null inputs are adjusted. For illustration purposes, consider Fig. 9.7 and assume that neuron 2 in the Kohonen layer is the winner with respect to an input stimulus $\{x\}$. In this circumstance, Fig. 9.7 illustrates those weights (continuous line arrows) in the Outstar layer that will be adjusted.

The adjustment rule applied in each one of these weights is synthesized by the following expression:

$$W_{ji}^{(2)} \leftarrow W_{ji}^{(2)} + \eta \cdot \left(d_j^{(k)} - W_{ji}^{(2)} \right) \cdot u_i^{(k)}, \tag{9.6}$$

where $u_i^{(k)}$ is the output of the kth neuron in the Kohonen layer, which won the competition (with respect to the kth training sample), $d_j^{(k)}$ is the kth component of the vector of desired outputs (with respect to the kth training sample), and η is the learning rate, which has the purpose of guiding the convergence process of the network.

For efficiency in the convergence, it is also suggested that the learning rate $\{\eta\}$ gets gradually decremented during the training process.

It can be observed that the adjustment rule presented in (9.6) performs an update in the weights of the quantizing neurons, which is proportional to the differences between the components of the weight vector and the components of the desired output vector. This principle has some similarity to Hebb's learning rule presented in Sect. 3.4. For the situation shown in Fig. 9.7, the weights that should be updated are the ones represented by $W_{1,2}^{(2)}$, $W_{2,2}^{(2)}$, and $W_{n_2,2}^{(2)}$.

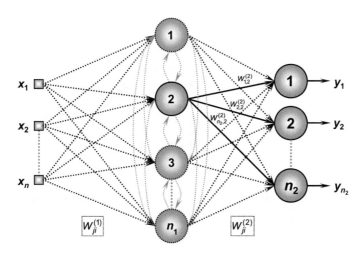

Fig. 9.7 Weight adjustment process of the outstar layer

From the analysis of the convergence process of the Outstar layer it can be verified that the weights of its neurons converge to the average values of the desired outputs, whereas the weights of the neurons in the Kohonen layer are adjusted with respect to the average of the input values (Wasserman 1989).

9.4.2   Training Algorithm of the Counter-Propagation Network

The training process of the counter-propagation network is divided into two stages. The first stage refers to the adjustments of the weights of neurons in the intermediate layer (Kohonen layer), which follows the same steps as the competitive algorithm presented in Sect. 8.2.

As to the second stage, it refers to the adjustments of the weights of neurons in the output layer (outstar layer) and assumes the update rule given by (9.6). This rule is applied only to those weights connected to the output of the winner neuron of the intermediate layer. In this second stage, the training process becomes supervised, since the use of desired outputs is required to tune the weight vectors.

The following algorithm shows, in pseudocode, the sequence of computational procedures for training a counter-propagation network.

BEGIN {COUNTER-PROPAGATION ALGORITHM – TRAINING PHASE}

<1> Obtain the set of training samples $\{x^{(k)}\}$;

<2> Obtain the desired output vector $\{d^{(k)}\}$ for each sample;

<3> Execute the training process of the intermediate (Kohonen) using the competitive algorithm // {according to Subsection (8.2)};

<4> Initialize $W_{ji}^{(2)}$ with small random values;

<5> Specify the learning rate $\{\eta\}$;

<6> Start the epoch counter {epoch ← 0};

<7> Repeat the instructions:

   <7.1> For all samples $\{x^{(k)}\}$, do:

      <7.1.1> Obtain the output values of the winner neuron of the intermediate layer (Kohonen) with respect to each sample $x^{(k)}$ // {according to Subsection (8.2)};

      <7.1.2> Adjust the individual weights of those neurons from the output layer (Outstar) that are connected to the winner neuron of the intermediate layer (Kohonen) // {according to equation (9.6)};

   <7.2> epoch ← epoch + 1;

   Until: there are no significant changes in the weight matrix $W_{ji}^{(2)}$.

END {COUNTER-PROPAGATION ALGORITHM – TRAINING PHASE}

After the training procedure, the network is ready to be put into operation when a new sample must be identified. The sequence of steps in the operation phase is given by the following algorithm.

BEGIN {COUNTER-PROPAGATION ALGORITHM – OPERATION PHASE}

<1> Present a sample $\{x\}$;

<2> Obtain the output values from the winner neuron of the intermediate layer (Kohonen) $\{u_i^{(v)}\}$ // {according to Subsection (8.2)};

<3> Calculate the network response from the outputs of the neurons in the the Outstar layer, using the following expression:

$$y_j = W_{ji}^{(2)} \cdot u_i^{(v)}, \text{ with } j = 1,...,n_2;$$

<4> Provide the network outputs given by the elements in y_j.

END {COUNTER-PROPAGATION ALGORITHM – OPERATION PHASE}

Furthermore, in accordance to the very creator of counter-propagation networks, Hecht-Nielsen (1987b), the counter-propagation network capacity to solve both pattern classification problems and curve fitting problems is very limited when compared to MLP and RBF networks. However, one of their advantages is on the simplicity of its computational implementation, being applied with success to several problems of engineering and science, especially those related to information and image compression (Dong et al. 2008; Mishra et al. 2006; Sygnowski and Macukow 1996).

9.5   Exercises

1. Explain the existing differences between data clustering and vector quantization.
2. Consider a class being identified by vector quantization. Explain in which situations the average values of the samples from this class could be adopted to allocate the corresponding quantizing vector.
3. Explain the main differences between MLP and LVQ networks when they are applied to pattern classification problems with respect to their decision boundaries.
4. Develop an algorithmic instruction that could be used as a stop criterion for the LVQ training process.
5. Explain the importance of initializing the weight vectors of the neurons of the LVQ with sample values of the corresponding classes they are representing.

Fig. 9.8 Spatial
representation of two classes

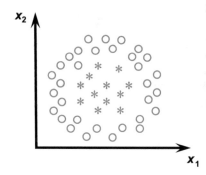

6. Consider two classes as shown in Fig. 9.8. Explain if the LVQ network can be trained to classify samples from these classes.
7. Explain the main differences between the training algorithms associated with LVQ-1 and LVQ-2 networks.
8. Explain the fundamental characteristics of both neural layers of the counter-propagation network architecture.
9. Explain whether it is possible to apply the generalized delta rule for adjusting the weights of neurons in the last layer of the counter-propagation network.
10. Comment about the reasons for not using conventional LVQ networks for curve fitting.

9.6 Practical Work

The forecast of the demand of electrical power is of utmost importance for the operational planning of the electrical power system. The maintenance schedule of the system, the expansion planning of the system and the dispatch analysis are usually performed by considering the foreseen electric power of the system.

The forecast for a particular day can be done by only taking into account the power measured in the early hours of the day. Based on this information, through a curve classification process, it is then possible to trace the profile of power demand necessary for the whole day.

Table 9.1 shows power measurements from 16 days grouped in four classes (demand profiles).

Therefore, construct and train a LVQ-1 network that detects possible similarities and regularities between all vectors that belong to each class, envisioning future classification of power profiles for other days (samples).

Table 9.1 Training data for the LVQ-1 network

Day	7:00	8:00	9:00	10:00	11:00	12:00	Class
1	2.3976	1.5328	1.9044	1.1937	2.4184	1.8649	1
2	2.3936	1.4804	1.9907	1.2732	2.2719	1.8110	1
3	2.2880	1.4585	1.9867	1.2451	2.3389	1.8099	1
4	2.2904	1.4766	1.8876	1.2706	2.2966	1.7744	1
5	1.1201	0.0587	1.3154	5.3783	3.1849	2.4276	2
6	0.9913	0.1524	1.2700	5.3808	3.0714	2.3331	2
7	1.0915	0.1881	1.1387	5.3701	3.2561	2.3383	2
8	1.0535	0.1229	1.2743	5.3226	3.0950	2.3193	2
9	1.4871	2.3448	0.9918	2.3160	1.6783	5.0850	3
10	1.3312	2.2553	0.9618	2.4702	1.7272	5.0645	3
11	1.3646	2.2945	1.0562	2.4763	1.8051	5.1470	3
12	1.4392	2.2296	1.1278	2.4230	1.7259	5.0876	3
13	2.9364	1.5233	4.6109	1.3160	4.2700	6.8749	4
14	2.9034	1.4640	4.6061	1.4598	4.2912	6.9142	4
15	3.0181	1.4918	4.7051	1.3521	4.2623	6.7966	4
16	2.9374	1.4896	4.7219	1.3977	4.1863	6.8336	4

Table 9.2 Testing data for the LVQ-1 network

Day	7:00	8:00	9:00	10:00	11:00	12:00	Class
1	2.9817	1.5656	4.8391	1.4311	4.1916	6.9718	
2	1.5537	2.2615	1.3169	2.5873	1.7570	5.0958	
3	1.2240	0.2445	1.3595	5.4192	3.2027	2.5675	
4	2.5828	1.5146	2.1119	1.2859	2.3414	1.8695	
5	2.4168	1.4857	1.8959	1.3013	2.4500	1.7868	
6	1.0604	0.2276	1.2806	5.4732	3.2133	2.4839	
7	1.5246	2.4254	1.1353	2.5325	1.7569	5.2640	
8	3.0565	1.6259	4.7743	1.3654	4.2904	6.9808	

For the simulation, consider a learning rate $\{\eta\}$ of 0.05. After the network training processed is finished, use it to classify power profiles for the days presented in Table 9.2.

Chapter 10
ART (Adaptive Resonance Theory) Networks

10.1 Introduction

The Adaptive Resonance Theory (ART), initially proposed by Grossberg (1976a, b), was developed from the observation of some biological phenomena, regarding vision, speech, cortical development, and cognitive-emotional interactions. This theory is based on three biological principles highlighted by the following characteristics:

- *Signal Normalization* → Ability of biological systems to adapt themselves to environments that change all the time. As an example, there is the human vision system, which can quickly adapt to different amounts of light.
- *Contrast Intensification* → Capability of identifying subtle details in the environment through successive observations. As an example, the respiratory system can differ, almost instantly, a clean environment that began being polluted with carbon monoxide.
- *Short-term memory* → Possibility to temporarily store sensorial information from the contrast intensification mechanism, before they can be decoded for decision-making.

Therefore, through this theory proposition, several artificial neural network models with unsupervised training (with recurrent characteristics) were sequentially proposed.

One of the main features of ART networks is the ability to learn new patterns, when new samples are presented, without destructing previously extracted knowledge. This characteristic is associated with the intriguing plasticity/stability dilemma (Carpenter and Grossberg 1988), in which the system must be flexible (adaptive) enough to incorporate environmental changes, whereas it must also be stable to preserve the knowledge already extracted during time.

Topologies based on the ART architecture can be divided into five major groups that are classified by input type and training process, this is:

© Springer International Publishing Switzerland 2017   189
I.N. da Silva et al., *Artificial Neural Networks*,
DOI 10.1007/978-3-319-43162-8_10

- *ART*-1 → Configuration with binary inputs and unsupervised training (Carpenter and Grossberg 1987a).
- *ART*-2 → Configuration with binary inputs or analogic inputs (continuous) and unsupervised training (Carpenter and Grossberg 1987b).
- *ART*-3 → Configuration with binary inputs or analogic inputs (continuous) and unsupervised training that uses multilevel topology and "chemical transmitters" for the searching process of a better solution (Carpenter and Grossberg 1990).
- *ART-Map* → Configuration with binary inputs or analogic inputs and supervised training in real time (Carpenter et al. 1991a) that uses two ART networks in its structure.
- *Fuzzy-ART* → Fuzzy version of the ART-MAP also with supervised training (Carpenter et al. 1991b).

Because of its high applicability, as well as its relatively simple implementation, in this book, only the ART-1 topology it is addressed, being the proprieties and principles related to the adaptive resonance theory the same to the five groups of ART networks.

Of course, as emphasized above, the essence of all ART topologies is based on the principle of incorporating knowledge from new input patterns, without substantially modifying the previous knowledge already inserted in the network from other input samples presented.

ART-1 network applications are as most diverse as possible, being found in several knowledge areas. As an example, in Young et al. (2007), an ART-1 architecture was used to detect spatial regions with astronomic interest, as well as, to characterize the noise existent in the collected information.

In Cho et al. (2006), a methodology was described, using ART-1, for computational grid-based classification aiming applications in bioinformatics. In Tateyama and Kawata (2004), a grouping technology using ART-1 networks was developed, to separate machines of an industrial plant into production cells for increasing productivity.

In Rajasekaran and Raj (2003) another ART-1 application was presented for feature extraction and recognition of monochromatic satellite images. In Baraldi and Alpaydin (2002), a neural structure based on the ART-1 network was also proposed for real-time data clustering.

Finally, in Srinivasan and Batur (1994) an integrated approach to Hopfield and ART-1 networks were presented for detection and fault isolation in linear dynamic systems.

10.2   Topological Structure of the ART-1 Network

The topological structure that represents an ART-1 network, designed by Carpenter and Grossberg (1987a), is illustrated in Fig. 10.1, where the input samples (vectors) are typically composed of binary values.

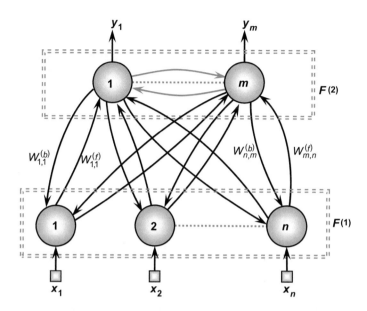

Fig. 10.1 Topological structure of the ART-1 network

The main applicability of ART-1 networks is in the solution of pattern classification problems, also being considered as a vector classification system.

In summary, by considering a sample represented by the input vector $\{x\}$, its purpose consists in classifying the sample into one of the classes in the problem, using for that a similarity measure between the sample vector and the other vectors that already represent those classes. It is worth noting that in the case of unsupervised learning, there are no desired outputs.

In addition, it can be observed in Fig. 10.1 that the ART-1 network consists of two layers of neuron ($F^{(1)}$ and $F^{(2)}$) that are totally connected to each another, this is, all neurons from layer $F^{(1)}$ are connected to all neurons from layer $F^{(2)}$ and vice versa.

The layer $F^{(1)}$ is denominated as the input layer or comparison layer and has the number of neurons equal to the dimensions of the input sample (dimension of x). Its purpose is also to pass original pattern x to the neurons in layer $F^{(2)}$. Therefore, as represented in Fig. 10.2, the flow of information inputted in layer $F^{(1)}$ is forwarded continuously (feedforward), this is, is always toward layer $F^{(2)}$.

As a consequence, the information fed back from layer $F^{(2)}$ are its input signals, composed of binary elements, which are weighted by its weight matrix (defined by $W^{(b)}$).

Hence, the primary role of layer $F^{(1)}$ is to perform a comparison to verify whether the input vector $\{x\}$, which is first associated to one of the classes represented by the neurons in the output layer, has similarity with the set of vectors already stored therein.

Fig. 10.2 Information flow
in ART-1 networks

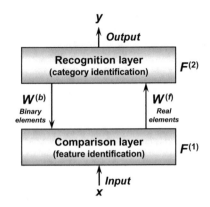

The layer $F^{(2)}$, denominated the output layer or the recognition layer, has the role of recognizing and incorporating the knowledge brought by an input vector to one of the classes represented by its neurons. Therefore, the number of neurons in this layer is dynamic, which varies according to the need of adding new neurons, so that new classes could be represented. This dynamic characteristic is one of the major potentials of ART networks.

Another feature of layer $F^{(2)}$ that is different from the behavior of input layer $F^{(1)}$ is that the neurons are competitive with each other (they have lateral connections). The flow of information in layer $F^{(2)}$ feeds back the neurons in the input layer $F^{(1)}$. The weight matrix of $F^{(1)}$ (defined as $W^{(f)}$) consists of real elements and is responsible for weighting the characteristics of the presented input vector to stimulate and activate the neurons representing the classes as a recognition signal.

Similarly to Kohonen networks, only a single neuron from the ART-1 recognition layer is active in each cycle, which is the winner of the proximity competition.

10.3   Adaptive Resonance Principle

The operation of the ART-1 network, when a new vector (sample) is presented to the input layer, is based on the adaptive resonance principle.

At first, each vector $\{x\}$ is presented and propagated from the input layer $F^{(1)}$ to the output layer $F^{(2)}$, where the component values of the vector are weighted by the corresponding weight matrix $\{W^{(f)}\}$ of the $F^{(2)}$ layer. This mechanism has the purpose of extracting discriminating characteristics that stimulate one of the classes represented by the neurons in $F^{(2)}$.

Second, when the pattern arrives (weighted by $W^{(f)}$) to the output layer $F^{(2)}$, the neurons in this layer, which are representing the classes of the problem being analyzed, begin a competition to verify which one has the greatest proximity with the received vector. A winner of the competition is then declared.

Third, the winner neuron feeds back the winner class to the input layer $F^{(1)}$. During this process, the neurons in the $F^{(1)}$ layer, using the weights stored in matrix $W^{(b)}$, perform a similarity test to verify the adhesion of the input vector to the class represented by the winner neuron in $F^{(2)}$.

Fourth, from the similarity test result, if the weight vector of the winner neuron is "similar" to the input vector, when considering an adhesion threshold (vigilance parameter), then there is resonance process between them. As a consequence, the input vector is then associated with the class that the winner neuron is representing. Therefore, the end of this recognition cycle is performed by the update of both weight matrices.

Although the current winner neuron (active) has won the competition against other neurons in $F^{(2)}$, if the input vector is not similar to the winner, then the neuron is not the appropriate one to represent the class associated with the entry. In this situation, in the next cycle, the same must be disabled so that other neurons in layer $F^{(2)}$ could have chances of winning the competition, and pass the new winning class to the similarity test. This process repeats itself for each new cycle until it is defined a class compatible with the input vector.

Finally, if none of the neurons in $F^{(2)}$ are approved in the similarity test, a new neuron must then be added/enabled in this layer, for the purpose of identifying this other class being presented by the input vector. In fact, this class is considered as a new class, since the other existing ones are not able to include new knowledge of features or behaviors brought by this input vector.

In fact, the resonance process performs a hypothesis test. In this circumstance, a hypothesis about an appropriate class that incorporates the input vector is formulated by the recognition layer $F^{(2)}$. Then the comparison layer $F^{(1)}$ comes into operation to validate the proposed hypothesis, this is, if the input vector has an acceptable degree of similarity to the suggested class, then the hypothesis is validated. Otherwise, the formulated hypothesis is considered wrong, and other classes (neurons) must be evaluated or even created in next cycles.

10.4   Learning Aspects of the ART-1 Network

The learning of new knowledge by the ART-1 network can be better understood using the flowchart systematized in Fig. 10.3.

Based on the flowchart blocks, it is possible to split the learning process of the ART-1 network into six main stages, which coincide with the shaded blocks in Fig. 10.3.

(I) First Stage—Parameter initialization phase

The instructions that make up this stage encompasses value assignment to the weight matrices $W^{(f)}$ and $W^{(b)}$, and the specification of the vigilance parameter $\{\rho\}$,

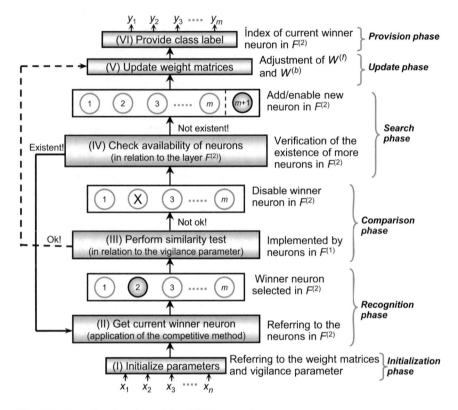

Fig. 10.3 Operation flowchart of the ART-1 network

whose role is to arbitrate the similarity test carried out by neurons in the input layer $F^{(1)}$.

The elements in weight matrix $W^{(f)}$ that are related to the output layer $F^{(2)}$ must be initialized with small values to avoid saturation of the results produced by neurons of this layer when an input vector is presented. Thus, elements in $W^{(f)}$ have the following values:

$$W_{ji}^{(f)} = \frac{1}{1+n}, \quad \text{with} \begin{cases} j = 1,\ldots,m \\ i = 1,\ldots,n \end{cases}, \tag{10.1}$$

where n is the number of neurons in the input layer $F^{(1)}$ and m is the number of enabled neurons in the output layer $F^{(2)}$.

The elements in weight matrix $W^{(b)}$ of the input layer $F^{(1)}$ are all initialized with unitary values, which indicate that each neuron of the layer $F^{(2)}$ are enabled to send its answer to the neurons in the $F^{(1)}$ layer. Thus:

$$W_{ij}^{(b)} = 1, \quad \text{with} \begin{cases} i = 1, \ldots, n \\ i = 1, \ldots, m \end{cases} \tag{10.2}$$

The vigilance parameter $\{\rho\}$ must have its value between 0 and 1 ($0 < \rho < 1$). It is also valid to observe that the bigger the ρ value, the more discriminating details are considered in the similarity test done by neurons in the layer $F^{(1)}$. This strategy allows that subtle differences between input patterns result in the creation of new classes, this is, more neurons are required to be enabled or added in the layer $F^{(2)}$.

On the other hand, very low values of the vigilance parameter result in a relaxation of the discriminating capacity of the network, since the same would evaluate only structural differences in the samples and, in this condition, the number of classes will be small.

(II) Second Stage—Recognition phase

The recognition stage consists in trying to categorize (frame) an input vector within those available classes, which are being represented by the neurons in the output layer $F^{(2)}$.

This recognition process is similar to the competitive method used in the Kohonen network, this is, the neuron from layer $F^{(2)}$ which has the biggest activation with respect to an input vector $\{x\}$ is then declared the winner of the competition. Therefore, the winner-takes-all strategy is applied, as explained in Sect. 8.2. The activation level of the jth neuron in the $F^{(2)}$ layer is produced by the signals of the input vector weighted by its corresponding weights. In mathematical terms:

$$u_j = \sum_{i=1}^{n} W_{ji}^{(f)} \cdot x_i, \quad \text{with } j = 1, \ldots, m \tag{10.3}$$

Consequently, the winner neuron $\{k\}$ is the one that produces the biggest activation value, this is:

$$k = \arg \max_j \{u_j\}, \tag{10.4}$$

where k is an integer value that represents the index of the current winner neuron (active), which in turn, inhibits the action of other neurons in the layer $F^{(2)}$ using its lateral connections. In this case, the winner neuron is representing the candidate class in which the input vector could be classified.

Additionally, the winner neuron produces a value equal to 1 in its output, which has the purpose to both indicate the candidate class of the input vector, as well as, feedback, in an excitatory way, the neurons in the layer $F^{(1)}$. On the other hand, by using lateral connections, all other neurons in the layer $F^{(2)}$ produce values equal to 0, indicating that the classes they are representing are not appropriate to incorporate the input vector, resulting, therefore, in an inhibitory feedback to the neurons in the layer $F^{(1)}$. In mathematical terms, we have:

$$y_j = \begin{cases} 1, & \text{if } j = k \\ 0, & \text{if } j \neq k \end{cases}, \quad \text{with } j = 1, \ldots, m, \tag{10.5}$$

where k is the index of the winner neuron.

Because the number of neurons in layer $F^{(2)}$ is dynamic, since more neurons are enabled or added as new classes are required, it can be stated that, when the first input vector is presented, one single neuron is available or enabled in layer $F^{(2)}$.

(III) Third Stage—Comparison Phase

After the procedure of categorizing the input vector during the recognition phase, the comparison phase is issued, whose purpose is to verify if the candidate class represented by the winner neuron is the appropriate one to receive the input vector.

The results produced by neurons in layer $F^{(2)}$ are fed back to the neurons in layer $F^{(1)}$, in order to accomplish the similarity test. As only the winner neuron k produces a value equal to 1 in its output, the fed back values from layer $F^{(2)}$ to layer $F^{(1)}$ are the synaptic weights associated with the connections between the winner neuron k and the neurons in $F^{(1)}$. Figure 10.4 illustrates this feedback process between neurons in layers $F^{(2)}$ and $F^{(1)}$.

The similarity test is, therefore, made by comparing the values of $W_{1,k}^{(b)}$ with x_1, of $W_{2,k}^{(b)}$ with x_2, and so on. As defined before, elements in matrix $W^{(b)}$ and elements in the vector x are binary, so to ease the comparison process. The most commonly used rule consists in comparing the vigilance parameter with the ratio R obtained by

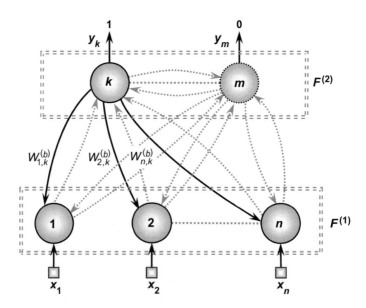

Fig. 10.4 Feedback from the winner neuron to the inputs

the number of common unitary elements (considering $W_{jk}^{(b)}$ and x_j) divided by the number of unitary elements in the vector x. This similarity ratio can be formally described by:

$$R = \frac{\sum_{j=1}^{n} W_{jk}^{(b)} \cdot x_j}{\sum_{j=1}^{n} x_j}, \qquad (10.6)$$

where k is the index of the winner neuron and n is the number of components in the vector x. As an example, assume that vector x is made of 8 input components ($n = 8$) and that the neuron with index 1 ($k = 1$) is the winner of the proximity competition. The following values are used for vectors x and $W^{(b)}$, as follows:

$$x_j = \begin{bmatrix} \overset{j=1}{1} & \overset{j=2}{0} & \overset{j=3}{1} & \overset{j=4}{1} & \overset{j=5}{0} & \overset{j=6}{1} & \overset{j=7}{0} & \overset{j=8}{1} \end{bmatrix}^T \qquad (10.7)$$

$$W_{j,1}^{(b)} = \begin{bmatrix} \overset{j=1}{1} & \overset{j=2}{0} & \overset{j=3}{1} & \overset{j=4}{0} & \overset{j=5}{0} & \overset{j=6}{1} & \overset{j=7}{1} & \overset{j=8}{1} \end{bmatrix}^T \qquad (10.8)$$

For this particular case, the numerator value in (10.6) is equal to 4, since the operation verifies just how many unitary elements are common to both vectors given in (10.7) and (10.8), which positions are $j = 1, j = 3, j = 6$, and $j = 8$. This operation results in a metric inspired by the Hamming distance (Moon 2005), which computes the quantity of different bits between two binary strings instead of computing the number of common bits. As for the denominator in (10.6), it indicates how many unitary elements are composing vector x, which have a return value of 5, when (10.8) is applied. Consequently, in this situation, the resulting value for the ratio R is equal to 0.8.

Therefore, if this ratio results bigger than the specified value for the vigilance parameter ($R > \rho$), then the input vector is approved in the similarity test. The class represented by the winner neuron k of $F^{(2)}$ is, in fact, the most appropriate to represent the characteristics of the input pattern. After that, an update in matrices $W^{(f)}$ and $W^{(b)}$ is done by applying Phase (V) to characterize the features from this new input vector.

Otherwise, if the ratio R results equal or smaller than the vigilance parameter ($R \leq \rho$), then it is assumed that the class represented by the winner neuron k is incapable of representing the behavior of the input vector. Therefore, there are significant differences between the input vector and those that are in the group of neuron k. Consequently, another neuron in the layer $F^{(2)}$ is selected as the winner and will have the chance to show that its class has more similarity with the sample vector.

(IV) Fourth Stage—Search Phase

In this processing stage, due to the winner neuron rejection in the similarity test, a search for another neuron (belonging to $F^{(2)}$) that may have more affinity with the sample is done. In this way, the winner neuron from the previous stage must be disabled to avoid that it could, again, win the competition. One of the ways to accomplish this operation is by assigning a null value to the activation level $\{u_k\}$ of that winner neuron so that another neuron could now win the competition.

Thus, as it can be observed in the flowchart of Fig. 10.3, subsequent sequential applications of phases (II), (III), and (IV) are done until it is possible to obtain a winner neuron in $F^{(2)}$ that is approved in the similarity test. Otherwise, if all neurons in the $F^{(2)}$ layer are reproved in the similarity test, then it is assumed that the input vector is representing a new class, since no existing neuron was capable of representing the features of the input. Consequently, a new neuron is added to layer $F^{(2)}$ in order to represent the features of that pattern. In fact, this new neuron perfectly incorporates the characteristics of the input in its weight vector, since this is the first sample represented by it.

It must be stressed that this ability of ART networks to incorporate, in an automatic way, new neurons in its output layer has been one of its main features, since the network becomes flexible enough to insert new knowledge in its structure without destroying information that already have been learnt. The algorithmic procedures for adding a new neuron in layer $F^{(2)}$ can be done in two ways.

The first way consists in using static data structures, such as a vector containing attributes from the neurons and with its maximum dimension assigned previously (the maximum number of neurons $F^{(2)}$ can have). In this circumstance, one of the attributes should report the neuron status, in a given cycle, indicating if the same is active (enabled) or inactive (disabled).

Alternatively, a counter that indicates the number of active neurons in layer $F^{(2)}$, in each cycle, could also be defined. As an example, if the maximum number of neurons in $F^{(2)}$ is defined 50, the counter would necessarily indicate that only a single neuron is active when the first sample is presented. This number is increased automatically as new neurons are enabled to represent other classes. Figure 10.5 illustrates this structural layout, taking into account that five neurons are currently active to represent classes, this is, only this five neuron could be considered for the proximity competition.

For the hypothetical case shown in Fig. 10.5, it can be observed that neuron number 4 has won the proximity competition, since it has the biggest activation value $\{u_k\}$, and, in this situation, only its output $\{y_k\}$ has a value equal to one. If this neuron and other neurons activated subsequentially are not approved in the similarity test, then neuron 6 is enabled through its status modification to "active."

The second way to add a new neuron in layer $F^{(2)}$ consists in the use of dynamic data structures, such as a linked list (Wirth 1989) containing each one of the attributes of the neurons. For this data structure, each neuron is considered an object interconnected by pointers, in order to make the structure of the logical list.

Neuron	1	2	3	4	5	6	7	...	50
Status	Active	Active	Active	Active	Active	Inactive	Inactive		Inactive
u_k	0.32	0.87	0.45	0.93	0.15	---	-----		-----
y_k	0	0	0	1	0	---	-----		-----

Fig. 10.5 Representation of neurons using static data structure

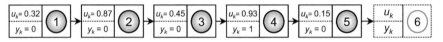

Fig. 10.6 Representation of neurons using dynamic data structure

So, initially, a single neural object is available while other neural objects are created and linked dynamically as new classes are required in layer F2. Therefore, differently from the static data structure, the status parameter is unnecessary, since all neurons from the list are active from the moment of their inclusion. Figure 10.6 illustrates this dynamic data structure that can also be used to represent neurons of the ART-1 network.

It can be noticed from Fig. 10.6 that when a new class is required, a new neuronal object (number 6) is created and incorporated into the previous dynamic structure. In this way, this structure has the characteristic of growing when it is needed.

(V) Fifth Stage—Update Phase

After a successful completion of the comparison phase (III), the update phase of matrices $W^{(f)}$ and $W^{(b)}$ proceeds to incorporate the knowledge brought by the last presented sample. For this purpose, only the elements concerning the winner neuron $\{k\}$ are updated by the following expressions:

$$W_{jk}^{(b)}(t+1) = W_{jk}^{(b)}(t) \cdot x_j, \quad \text{with } j = 1, \ldots, n \tag{10.9}$$

$$W_{kj}^{(f)}(t+1) = \frac{W_{jk}^{(b)}(t) \cdot x_j}{\frac{1}{2} + \sum_{i=1}^{n} W_{ik}^{(b)}(t) \cdot x_i}, \quad \text{with } j = 1, \ldots, n \tag{10.10}$$

The weight update of the winner neuron with respect to matrix $W^{(b)}$, given in (10.9) and seen in Fig. 10.4, is processed using a simple Boolean logic operation

Table 10.1 Example of adjustment values for $\mathbf{w}^{(b)}$

		$j=1$	$j=2$	$j=3$	$j=4$	$j=5$	$j=6$	$j=7$	$j=8$
Winner weights	$W_{j,2}^{(b)}(t)$	1	0	0	1	1	1	0	1
ART-1 inputs	x_j	1	0	1	1	1	0	1	1
$W^{(b)}$ adjustment (AND gate)	$W_{j,2}^{(b)}(t+1)$	1	0	0	1	1	0	0	1

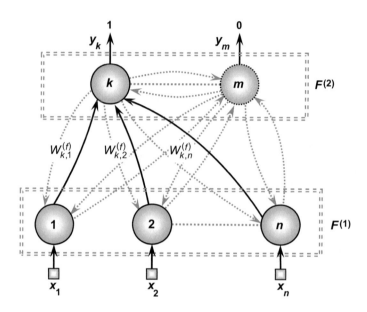

Fig. 10.7 Flow of the input to the winner neuron

(AND gate) between the elements of $W_{jk}^{(b)}$ and x_j. In Table 10.1 it is shown the result of expression (10.9) when it is applied to a hypothetical ART-1 network made of 8 inputs $\{n = 8\}$, and when it is assumed that the winner neuron with $\{k = 2\}$ is the one approved in the similarity test.

The weight adjustment of the winner neuron with respect to matrix $\mathbf{W}^{(f)}$, in which the connections of interest are highlighted in Fig. 10.7, is done by two operations.

The first operation, done by the numerator of (10.10), performs the application of an AND gate given by (10.9). For the second operation, done by the denominator of (10.10), the resulting summation is the number of unitary elements that are

Table 10.2 Adjustment values for $W^{(f)}$

		$j = 1$	$j = 2$	$j = 3$	$j = 4$	$j = 5$	$j = 6$	$j = 7$	$j = 8$
Winner weights	$W_{j,2}^{(b)}(t)$	1	0	0	1	1	1	0	1
ART-1 inputs	x_j	1	0	1	1	1	0	1	1
$W^{(f)}$ adjustment	$W_{2,j}^{(f)}(t+1)$	$\frac{2}{9}$	0	0	$\frac{2}{9}$	$\frac{2}{9}$	0	0	$\frac{2}{9}$

juxtaposed in the weights of the winner neuron ($W^{(b)}$) and in the components of the input vector.

In Table 10.2, a numerical example of the $W^{(f)}$ adjustment is illustrated, considering the same conditions and parameters used to obtain Table 10.1.

In this example, it can be verified that there is an excitatory stimulus in $W_{2,j}^{(f)}(t+1)$, only for those connections whose components $W_{j,2}^{(b)}(t)$ and x_j have concomitantly unitary values.

(VI)   Sixth Stage—Provision Phase

According to the flowchart illustrated in Fig. 10.3, the sixth stage in the training process of the ART-1 is to provide the label of the class in which the sample will be assigned. In this case, after performing all previous stages, it is then enough to apply expression given in (10.5) to obtain the selected class, i.e., the class represented by the winner of the proximity competition.

10.5   Training Algorithm of the ART-1 Network

The training process of the ART-1 network is structured from the six stages shown in the previous section, which the instructions in the pseudo code are described next.

Unlike other neural architectures studied throughout this book, it is verified that both the training phase and the operating phase of the ART-1 network are included in the same algorithm, since this topology always needs to perform a similarity test to categorize the input sample. Furthermore, this learning is processed in an unsupervised manner, allowing the inclusion of knowledge inside classes that are already represented by existing neurons, or, evaluating if there is a necessity for the inclusion or enablement of other neurons as the samples bring new relevant knowledge.

BEGIN { ART-1 ALGORITHM – TRAINING AND OPERATION PHASE}

<1> Present the input vector {x};

<2> Initialize matrices $W^{(f)}$ and $W^{(b)}$; {according to (10.1) and (10.2)}

<3> Specify the vigilance parameter ρ; {defined between 0 and 1}

<4> Repeat the instructions:

 <4.1> Calculate neural activations {u_j}; {according to (10.3)}

 <4.2> Obtain the current winner neuron {k}; {according to (10.4)}

 <4.3> Calculate the similarity ratio {R}; {according to (10.6)}

 <4.4> If ($R > \rho$) do:

  <4.4.1> Update $W^{(b)}$ and $W^{(f)}$; {according to (10.9) and (10.10)}

  <4.4.2> Assign x to the class of the winner neuron;

  <4.4.3> Winner ← "Approved";

 <4.5> If {($R \leq \rho$) and (Available neurons exist)} do:

  <4.5.1> Disable current winner neuron {$u_k = 0$};

  <4.5.2> Winner ← "Reproved";

 <4.6> If {($R \leq \rho$) and (Available neurons do not exist)} do:

  <4.6.1> Include or enable a new neuron;

  <4.6.2> Update $W^{(b)}$ and $W^{(f)}$; {according to (10.9) and (10.10)}

  <4.6.3> Assign x to the class of the new neuron;

  <4.6.4> Winner ← "Approved";

 Until: Winner = "Approved";

<5> Enable again all neurons in $F^{(2)}$.

END {ART-1 ALGORITHM – TRAINING AND OPERATION PHASE}

In fact, the ART-1 network implements the clustering algorithm known as the leader-follower algorithm (Duda et al. 2001), which is also called follow-the-leader strategy (Hartigan 1975), whose primary purpose is to find an appropriate class that could group an input sample. If the sample has a high-level deviation from the elements stored in the candidate class, the algorithm objective becomes to create a new class to store the sample attributes.

10.6 Aspects of the ART-1 Original Version

The original version of the ART-1 network, as conceived by Carpenter and Grossberg (1987a), is yet formulated in terms of first-order differential equations, and also has three binary control or gain parameters (G_2, G_1, $Reset$). These parameters are included in the network algorithm to control the dynamic flow of information between layers $F^{(1)}$ and $F^{(2)}$, this is:

- *Parameter "G_2"* → Responsible for the status that activates the execution flow for layer $F^{(2)}$, and has the purpose of enabling the allocation of new neurons in the layer. This parameter output is active {1} if at least one component of the input vector {*x*} is equal to 1. This output can be visualized as the result of applying the logic gate "OR" in all components of the input vector.

- *Parameter "G₁"* → Responsible for the status that activates the execution flow for layer $F^{(1)}$. Its output is active {1} if at least one component of the input vector {*x*} is equal to 1 and all outputs of neurons in layer $F^{(2)}$ are null.
- *Parameter "Reset"* → Responsible for the status obtained by the similarity test. It provides a response in active state {1} if the winner neuron in $F^{(2)}$ fails the similarity test.

As seen in Fig. 10.8, each neuron in layer $F^{(1)}$ receives three binary values, i.e., the corresponding component of the input vector, the corresponding weighted summation from neurons of the output layer $F^{(2)}$, and the control parameter G_1, which is the same for all of them.

The execution flow is then in the $F^{(1)}$ layer when the completion of the "two-thirds" rule is observed, i.e., at least, two of these three values must be active (producing outputs equal to one). A better description of this mechanism is shown in Carpenter and Grossberg (1987b).

The operation of the original version of the ART-1 network, with its control parameters, follows the same instructions of the algorithm presented in Sect. 10.5, which can be synthesized by the following steps:

(i) Initially, when a sample is presented, all outputs of the neurons of $F^{(2)}$ are producing null values. In this situation, the active layer is layer $F^{(1)}$, since the "two-thirds" rule is verified as both G_1 and G_2 are producing values equal to 1.

(ii) One copy of the sample is then forwarded to layer $F^{(2)}$, in which it is defined the winner neuron. The winner neuron produces a unitary output value, and all other neurons produce null output values. At this time, due to the unitary

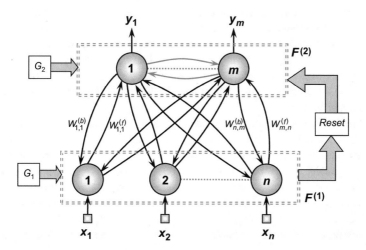

Fig. 10.8 Control parameters of the original version of the ART-1 network

output of the neuron, G_1 becomes zero {Similar to the application of (10.4) and (10.5)}.

(iii)   The execution flow returns to $F^{(1)}$. Then, only those neurons in $F^{(1)}$ that have components of x equal to 1, and those that received unitary values from the weighted outputs of the neurons of $F^{(2)}$ are activated (will produce unitary values on their outputs). This matched pattern $\{v\}$ is then sent to the similarity test.

(iv)   If the number of unitary components in both v and x are very close (bigger than the vigilance parameter) then, the reset value should be null and, consequently, an update in matrices $W^{(b)}$ and $W^{(f)}$ is performed {similar to the application of (10.6), (10.9), and (10.10)}. Otherwise, the reset parameter assumes unity, indicating that the current winner neuron must be disabled.

Therefore, it can be stated that these four operating steps are also integrally implemented in the algorithm presented in the previous section, i.e., the three control parameters are implicit within those conditions imposed by the comparison and repetition structures.

As it could be verified throughout this chapter, one capability of the ART-1 network is its flexibility to incorporate new knowledge, without destroying all that have been previously learnt. This distinctive quality makes it one of the best architectures of artificial neural networks, which can deal with the stability/plasticity dilemma in a coherent and systematic way. Complementary, some other distinctive characteristics are the following (Carpenter and Grossberg 1987a):

(i)   The ART architecture has biological plausibility, which is principled by the signal normalization, contrast intensification, and short-term memory principles.

(ii)   The network training is always stable and after this stabilization (convergence), the presentation of a pattern that fits one category already created, directly activates the neuron corresponding to that group, with no need to initiate the search phase. In this case, the network works as an autonomous associative memory.

(iii)   The selection of the winner neuron in the recognition phase is also always stable. Once an input vector is associated with a group represented by a neuron in $F^{(2)}$, this same unit always wins the competition, regardless of some eventual posterior adjustment in $W^{(f)}$ and $W^{(b)}$ when a new training sample is presented.

(iv)   The occurrence of adaptive resonance depends, mainly, on how close is the input sample to the vector that represents the cluster, indicated by the winner neuron. If the distance, weighted by the vigilance parameter $\{\rho\}$, is acceptable, then, an adaptive resonance state is achieved, what in biological terms, corresponds to the gain and extension of neural activity.

(v)   The level of details in each new class included in the network structure is based on the vigilance parameter value. The larger the value, the more fine details and distinctive characteristic are considered from the patterns to be inserted.

10.7 Exercises

1. Enumerate the main features regarding layers $F^{(1)}$ and $F^{(2)}$ of the ART-1 network.
2. Explain the differences between the Kohonen and the ART-1 architectures from a plasticity point of view.
3. What is the adaptive resonance principle and how it can be used to evaluate ART-1 networks.
4. Briefly, explain how the acquisition of new knowledge is processed in ART-1 networks.
5. Write about the plasticity/stability dilemma when a multilayer perceptron is analyzed.
6. Explain which computational data structures can be used to implement the addition or enablement of new neurons in the output layer of an ART-1 network.
7. Explain what is the main purpose of the vigilance parameter in an ART-1 network.
8. Regarding the answer to the previous question, explain what are the consequences when assuming a unitary value for the vigilance parameter.
9. Let an input vector be given by $\mathbf{x} = [1\ 0\ 1\ 1\ 1\ 0\ 1\ 0]^T$. Consider that the neuron of index k is the winner of layer $F^{(2)}$, then the elements associated to the weights of neurons in $F^{(1)}$ are:

$$W_{j,k}^{(b)} = [1\ \ 1\ \ 0\ \ 0\ \ 1\ \ 1\ \ 1\ \ 0]^T$$

Write about whether the pattern would be approved in the similarity test when a value of 0.8 is considered for the vigilance parameter.
10. Regarding the previous exercise, what should be the minimum value assumed for the vigilance parameter so that the pattern can be approved in the similarity test.

10.8 Practical Work

The behavior of an industrial process can be analyzed by taking into consideration several variables related to the phases of the process. In Table 10.3 there are 10 different situations of the system behavior from 16 status variables $\{x_1, x_2, \ldots, x_{16}\}$ collected.

Hence, the purpose of this project is to implement and train an ART-1 network that classifies and groups similar situations into classes, so that when an eventual requirement of maintenance is issued, one could have a probable diagnosis.

Table 10.3 Training set for the ART-1 network

	x_1	x_2	x_3	x_4	x_5	x_6	x_7	x_8	x_9	x_{10}	x_{11}	x_{12}	x_{13}	x_{14}	x_{15}	x_{16}
Situation 1	0	1	0	1	1	0	1	0	1	1	0	1	1	1	1	1
Situation 2	1	0	1	0	1	1	1	1	1	1	1	0	1	0	0	0
Situation 3	1	0	1	1	1	1	1	0	1	1	0	1	1	0	1	1
Situation 4	1	1	1	0	1	0	1	0	1	1	1	1	0	1	0	0
Situation 5	0	0	1	1	1	1	1	1	0	1	1	0	0	0	0	1
Situation 6	1	1	0	1	0	0	1	0	1	1	0	1	1	1	1	1
Situação 7	1	0	1	0	1	1	0	1	1	1	1	0	1	1	1	0
Situação 8	1	0	1	1	1	1	1	0	1	1	0	1	1	0	1	1
Situação 9	0	1	1	0	1	0	1	0	1	1	0	1	0	1	0	1
Situação 10	0	0	1	1	1	1	1	1	0	1	1	0	0	0	0	1

Proceed with the simulations for this classification, considering the following values for the vigilance parameter: (i) $\rho = 0.5$; (ii) $\rho = 0.8$; (iii) $\rho = 0.9$; and (iv) $\rho = 0.99$.

Finally, after each simulation, indicate the number of active classes, through their corresponding neurons, and list which situations in Table 10.3 are being included in its correct groupings.

Part II
Application of Artificial Neural Networks in Engineering and Applied Science Problems

Chapter 11
Coffee Global Quality Estimation Using Multilayer Perceptron

11.1 Introduction

This application uses artificial neural networks for qualifying coffee batches or coffee brands from a set of sensors based on conductive polymers, which were developed by EMBRAPA (Brazilian Agricultural Research Corporation). The set, known as the "electronic tongue," is shown in Fig. 11.1.

This device enables a quality characterization of the analyzed coffee, which is something traditionally done by professional tasters by giving the sample a graded score between 1 and 10.

The electronic tongue performs capacitive reactance measurements in the samples (in a liquid environment) using 55 distinct frequencies. Using artificial neural networks, it is desired to extract, from this information, all those features that could be used for creating a model capable of assigning grades to coffee batches or coffee brands.

Therefore, this problem consists basically in fitting a curve whose domain is defined by frequencies from the operation of the electronic tongue, which gives a measure of the coffee quality as a response. The response is tabulated on a scale from 1 to 10. These values are related to those assigned by specialized tasters, being the score ten to the maximum quality level.

11.2 MLP Network Characteristics

The input signals were normalized between [0; 1] for a better output conditioning. The ANN used in this application was a multilayer perceptron (MLP).

Several MLP topologies were tested, finally leading to two candidates with good results, this is, Topology T1 (with one hidden layer) and Topology T2 (with two hidden layers), whose configurations are shown as follows:

© Springer International Publishing Switzerland 2017
I.N. da Silva et al., *Artificial Neural Networks*,
DOI 10.1007/978-3-319-43162-8_11

Fig. 11.1 The "Electronic Tongue" set

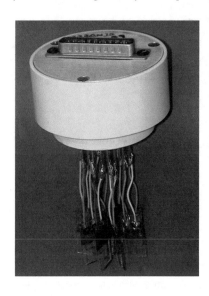

- **TOPOLOGY T1**

  – **Input Layer** → 55 input signals, related to the capacitance measurements in 55 different frequencies.
  – **Hidden Neural Layer** → 28 neurons.
  – **Output Neural Layer** → 1 neuron.

- **TOPOLOGY T2**

  – **Input Layer** → 55 input signals, related to the capacitance measurements in 55 different frequencies.
  – **First Hidden Neural Layer** → 10 neurons.
  – **Second Hidden Neural Layer** → 2 neurons.
  – **Output Neural Layer** → 1 neuron.

The Backpropagation algorithm was used as the training algorithm in the learning phase. Cross-validation with a dataset made up of 236 examples was used in the first place. From the data set, 90 % of the data were used for training and 10 % of the data for validation.

Next, a k-fold cross-validation was applied to increase the generalization ability of the network. Samples were divided into 10 subsets, having 6 subsets with 24 samples and 4 subsets with 23 samples.

The learning rate used was 0.1. The specified number of iterations for the network convergence was 500 epochs. The tested topologies can be better understood in Figs. 11.2 and 11.3.

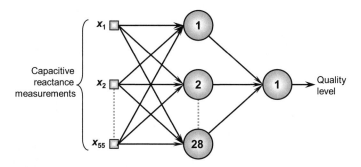

Fig. 11.2 Illustration of topology T1, with one hidden layer

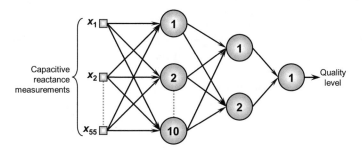

Fig. 11.3 Illustration of topology T2, with two hidden layers

11.3 Computational Results

Correlation factors (r^2) between the desired outputs and the outputs obtained by the networks are presented in Table 11.1. Here, results are obtained using the validation subsets.

In Fig. 11.4, it is shown a comparison between the curve fitting done by topology T1 and the desired values.

In Fig. 11.5 it is possible to see the comparative analysis between the curve fitting done by topology T2 and the desired values.

Both topologies tested were capable of maintaining the correlation factor (r^2) near 0.8. This result is considered promising if compared to grade assignments done by professional tasters.

Table 11.1 Training results

Validation method	Correlation factor (r^2) (Topology T1)	Correlation factor (r^2) (Topology T2)
Conventional cross-validation	0.8437	0.8648
K-fold cross-validation	0.7998	0.8184

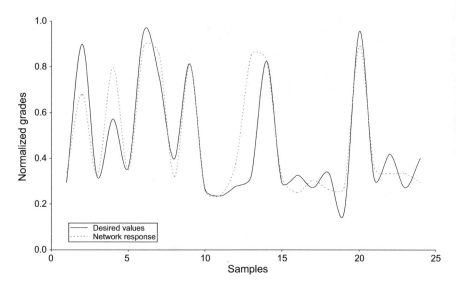

Fig. 11.4 Comparative results (topology T1)

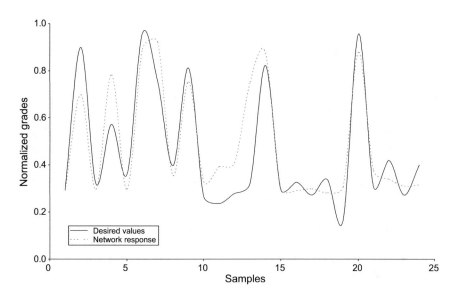

Fig. 11.5 Comparative results (topology T2)

Topology T1 had better results, and, because it has only one hidden layer, it has a better computational efficiency when compared to topology T2.

Hence, the developed model was able to perform the curve fitting by using the "Electronic Tongue," and it is then able to assist tasters in the process of assigning grades for coffee evaluation.

Chapter 12
Computer Network Traffic Analysis Using SNMP Protocol and LVQ Networks

12.1 Introduction

Investigations of data flow in a computer network are essential for planning the system expansion, as well as for solving problems.

The main difficulty in managing a computer network is to analyze results available from traffic monitoring, since data are not intuitive, and therefore, an expert inspection can be necessary.

Tests regarding traffic information in a computer network can be done in three ways:

- Analysis of all data transmissions done in the whole network or a particular segment of the network by an interface in promiscuous mode.
- Traffic analysis of a particular interface that belongs to a host (device or equipment) and to a computer, which is suspected to have a network traffic flooding.
- Statistical traffic analysis of devices or servers in the segment.

The SNMP (Simple Network Management Protocol) is a network management protocol that does an active network monitoring and provides equipment configuration services (Kurose and Ross 2016]. This protocol acts in the application layer of the OSI model.

Information is obtained with the SNMP through a request from the manager to one or more agents, as can be examined in Fig. 12.1.

Each machine in the network is represented as a set of variables related to its current state. Such information is available to the manager, and the latter can also modify it. Each managed machine must have an agent and an MIB database (*Management Information Base*).

The agent is a process running on the machine being managed while the manager is a process running on the server, which allows the communication between the manager and several agents being managed, as it can be observed in Fig. 12.2.

© Springer International Publishing Switzerland 2017
I.N. da Silva et al., *Artificial Neural Networks*,
DOI 10.1007/978-3-319-43162-8_12

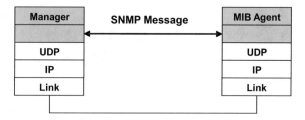

Fig. 12.1 Relationship between "manager" and "agent" based on the TCP/IP model

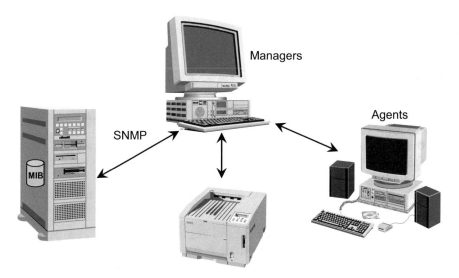

Fig. 12.2 Mechanism for a communication management between devices

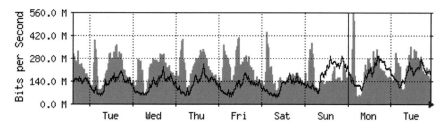

Fig. 12.3 Example of a traffic chart

The management of the internet link, which provides the transmission band, is often done visually by graphics provided by specialized tools, as illustrated in Fig. 12.3.

Therefore, the network classification, in a particular moment, must be performed by an expert. This fact could lead to delays in decision-making if some interpretation error happened.

Here, the application of LVQ networks in the link classification process was investigated in order to make the classification of a computer network automatic.

12.2 LVQ Network Characteristics

An LVQ-1 network does the automatic traffic classification. In Fig. 12.4, it can be seen the behavior of the analyzed network for a 1 week period; wherein output traffic was kept bellow 200 kbps (kilobit per second). In this investigation, for all studies on network traffic, a single Ethernet interface in a server was analyzed.

The LVQ-1 network topology was made of 4 prototype neurons, so to represent the following traffic classes:

- *Class A* → Traffic up to 43 kbps.
- *Class B* → Traffic between 43 and 86 kbps.
- *Class C* → Traffic between 86 and 129 kbps.
- *Class D* → Traffic over 129 kbps.

Figure 12.5 illustrates the LVQ-1 network structure, where the outputs represent the traffic classes.

A total of 36 laboratory samples were used for training the network, which represent 9 situations of traffic behavior regarding each one of the previously defined classes. The acquisition of bits from the network output was performed every 5 min for a 24 h interval, totalizing 265 input signals for the LVQ-1 network.

12.3 Computational Results

Ten different training processes (t1, t2, …, t10) were used. In each one of them, new network weights were generated. Based on the hits of the results, it was then possible to construct a chart with training performances for each day of the week, as exhibited in Fig. 12.6.

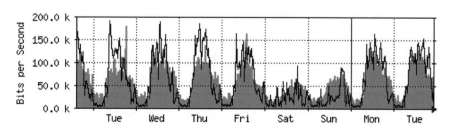

Fig. 12.4 Behavior of the analyzed network

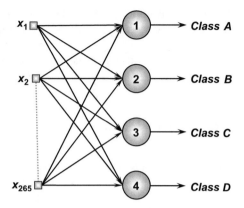

Fig. 12.5 LVQ-1 network topology for traffic classification

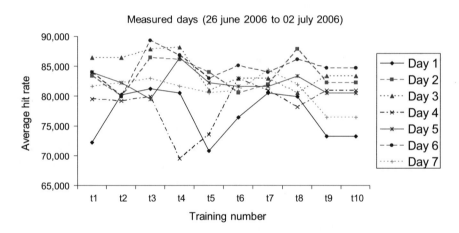

Fig. 12.6 Performance during 7 days

For classification, 288 patterns were used, each one also acquired in intervals of 5 min.

An average hit rate of 81.57 % could be observed, being the minimum rate equal to 69.44 %, and the maximum rate equal to 89.23 %. Results in Fig. 12.7 supports the applicability of the LVQ network for interpreting traffic classes and provides a report of the current network class to the user.

For this application, the great advantage of using an LVQ for classifying networks is the low computational cost, since results are compiled in a short space of time (less than half a second).

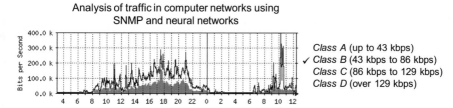

Fig. 12.7 Visualization of the proposed implementation

Additionally, hit rates obtained from the proposed approach are also considered favorable to analyzing traffic in computer networks.

Chapter 13
Forecast of Stock Market Trends Using Recurrent Networks

13.1 Introduction

Conventional methods for predicting the behavior of financial papers are based on specialist's analysis and decision-making, and automatic methods are unavailable for most situations.

Within this context, the use of recurrent artificial neural networks constitutes an alternative to decision-making in the financial stock market.

This application contributes to already-existent tools for decision-making in stock markets, since it analyzes the influence of macroeconomic variables.

In capital markets, there are two models that rule strategies of share sales and purchases. The fundamentalist model analyzes the economic and accounting aspects of companies as well as the influence of macroeconomic variables. As for the technical model, it is based on the historical data of a company, trying, therefore, to predict the future based on past behaviors.

Both models are not mutually exclusives, but complementaries, and any decision in the stock market is made by evaluating both studies.

In the financial market, several companies offer several types of shares for sale, such as PN and ON. The PN is a type of share that gives one the right of receiving profit from the profit of the company. The ON gives one the right of voting in the company meetings.

Therefore, the price history of a particular share produces a sequence of numbers whose behavior forecasting is essential for future deals. In Fig. 13.1, it can be observed the behavior of prices of the Pão de Açúcar Group in 2005.

In Fig. 13.1, the nonlinear behavior of the curve is evidenced and reveals the necessity of using tools capable of dealing with this type of problem in an automatic and efficient way.

Consequently, as it is desired to evaluate share prices in future periods, the most appropriate neural architecture is the one with recurrent configuration, which is capable of forecasting several steps ahead in the future from historical data.

© Springer International Publishing Switzerland 2017   221
I.N. da Silva et al., *Artificial Neural Networks*,
DOI 10.1007/978-3-319-43162-8_13

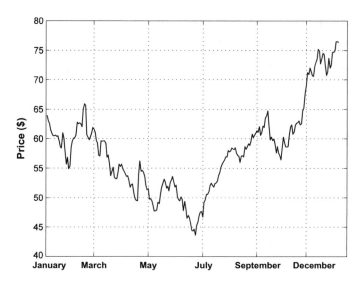

Fig. 13.1 Stock price evolution chart of the Pão de Açúcar Group in 2005

13.2 Recurrent Network Characteristics

The neural network used for estimating share values is a recurrent network (simple), also denominated Jordan network (Sect. 5.4.3), similar to the one shown in Fig. 13.2.
The neural network topology used is described as follows:

- **Input Layer** → 5 inputs, being them: the opening price $\{x_1\}$, the maximum price $\{x_2\}$, the minimum price $\{x_3\}$, the closing price $\{x_4\}$, and the commercialized volume $\{x_5\}$.
- **Hidden Neural Layer** → 10 neurons.
- **Output Neural Layer** → 1 neuron, representing the closing price of the shares for future days.

Fig. 13.2 Illustration of the recurrent neural network

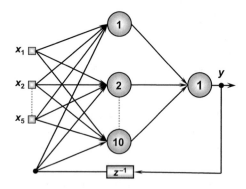

A forecast evaluation was done by taking into consideration not only historical factors but also economic variables that influence share prices. In this case, a second neural network that considers such variables was trained. Its topology is configured as follows:

- **Input Layer** → 8 inputs, being them: the opening price, the maximum price, the minimum price, the closing price, the commercialized volume, the short-term interest rate, the industrial production index, and the BOVESPA index.
- **Hidden Neural Layer** → 10 neurons.
- **Output Neural Layer** → 1 neuron, representing the closing price of the shares for future days.

13.3   Computational Results

For validating the efficiency of the artificial neural networks, 3 study cases were considered, regarding the behavior of the shares of the Pão de Açúcar Group, ITAUSA Group, and EMBRAER.

In Fig. 13.3 it is possible to see the results of estimating shares of the Pão de Açúcar group, taking into consideration only historical data of the shares prices.

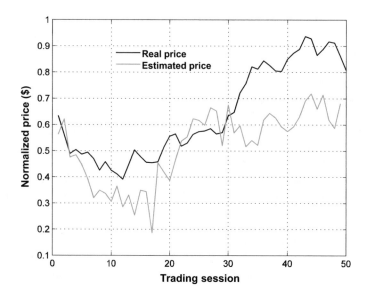

Fig. 13.3 Price estimation for the Pão de Açúcar Group using historical data

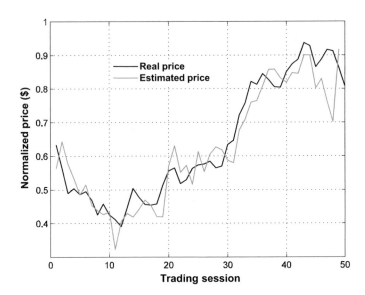

Fig. 13.4 Price estimation for the Pão de Açúcar Group using historical and economic data

In Fig. 13.4, the results of estimating shares of the Pão de Açúcar Group are taking into consideration the 3 macroeconomic variables described above. The forecast presents an improvement of 85 % with respect to the forecast done using only historical variables

The results using economic variables were significantly better; however, its use for long-term forecasts becomes inappropriate, because errors are forwarded as the forecasting horizons increases.

The results for the ITAUSA Group are in Figs. 13.5 and 13.6, where the forecast was also done with and without the influence of economic variables.

Once again, it can be noticed that the estimation has a better performance when economic variables are present in the inputs of the neural network.

The estimation results for EMBRAER are in Figs. 13.7 and 13.8.

As it can be seen in Figs. 13.7 and 13.8, the economic variables added to the recurrent network have little influence in enhancements on the estimation results of share prices of EMBRAER. Such behavior arises from the fact that this company is on the aeronautics business, which is designated for the foreign market, and internal economic variables affect the performance just slightly.

Therefore, for EMBRAER case, a bigger training database was used, resulting in significant enhancements in the estimation results as illustrated in Fig. 13.9.

These promising results showed the efficiency of the recurrent neural network in predicting time series with nonlinear behavior, such as those in the stock market.

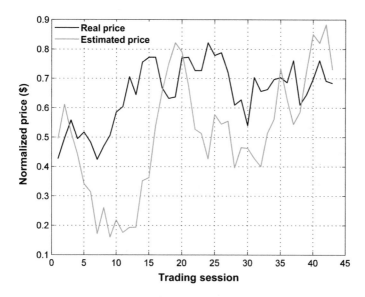

Fig. 13.5 Price estimation for the ITAUSA Group using historical data

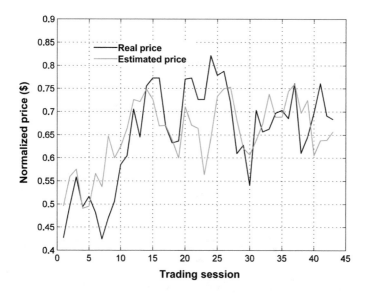

Fig. 13.6 Price estimation for the ITAUSA Group using historical and economic data

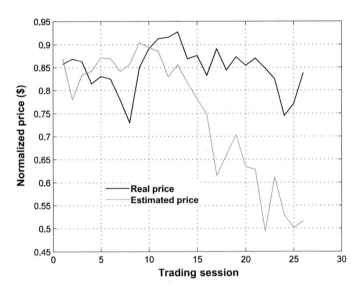

Fig. 13.7 Price estimation for EMBRAER using historical data

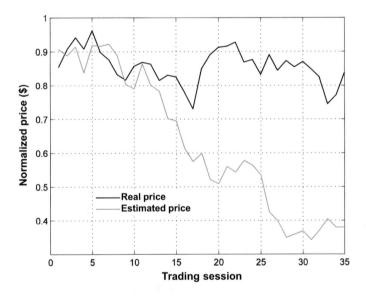

Fig. 13.8 Price estimation for EMBRAER using historical and economic data

The use of economic variables enabled a substantial enhancement on the estimation results, mainly in shares whose companies are heavily dependent on national financial indexes.

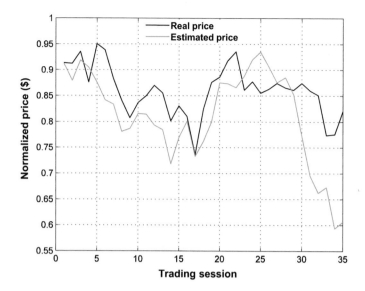

Fig. 13.9 Price estimation for EMBRAER using an incremented database

As for companies whose market is regulated by international influences, as the aeronautics market, such economic variables poorly affect the quality of the estimation. Therefore, it requires the study of other indexes capable to better sensitize the artificial neural networks.

Chapter 14
Disease Diagnostic System Using ART Networks

14.1 Introduction

Specification of the correct treatment for people with disorders can be very complex due to the diversity of disorders. To this end, this chapter presents an application of artificial neural networks for assisting a more precise identification of diagnostics regarding such diseases.

According to ICD-10 (International Statistical Classification of Diseases and Related Health Problems), disturbances are divided into 21 classes as follows:

1. Certain infectious and parasitic diseases.
2. Neoplasms (tumors).
3. Diseases of the blood and blood-forming organs, and certain disorders involving the immune system.
4. Endocrine, nutritional, and metabolic diseases.
5. Mental and behavioral disorders.
6. Diseases of the nervous system.
7. Diseases of the eye and adnexa.
8. Diseases of the ear and mastoid process.
9. Diseases of the circulatory system.
10. Diseases of the respiratory system.
11. Diseases of the digestive system.
12. Diseases of the skin and subcutaneous tissue.
13. Diseases of the musculoskeletal system and connective tissue.
14. Diseases of the genitourinary system.
15. Pregnancy, childbirth, and puerperium.
16. Certain conditions originating in the perinatal period.
17. Congenital malformations, deformations, and chromosomal abnormalities.
18. Symptoms, signs and abnormal clinical, and laboratory findings, not elsewhere classified.

© Springer International Publishing Switzerland 2017
I.N. da Silva et al., *Artificial Neural Networks*,
DOI 10.1007/978-3-319-43162-8_14

Fig. 14.1 Supporting
diagnostic system of a patient

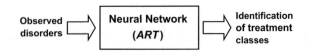

Observed disorders ⟹ Neural Network (ART) ⟹ Identification of treatment classes

19. Injury, poisoning, and certain other consequences of external causes.
20. External causes of morbidity and mortality.
21. Factors influencing health status and contact with health services.

The presence of two or more problems makes difficult the choice of treatments to a patient. Therefore, for n patients, depending on current conditions, several types of treatment could be necessary. A system based on the ART (*Adaptive Resonance Theory*) artificial neural network is projected, as presented in Chap. 10, to assist the identification of how many classes of treatments are needed for a certain number of patients.

As illustrated in Fig. 14.1, through the information of the disorders on n patients, it is then possible to point out how many treatment classes are required.

In this application, the ART neural network is used, as it will allow a better definition of classes of treatment necessary for a certain number of patients.

14.2 Art Network Characteristics

Table 14.1 presents indexes for the classes of the diseases, as classified by ICD. For a particular patient, a value of 1 (one) was considered for the manifestation of a disease and a value of 0 (zero) was considered for the absence of a disease.

As a consequence, by presenting this information to the ART network, the same should be able to identify the number m of different classes of treatment required for a set of patients being evaluated. Figure 14.2 illustrates the ART network topological configuration used in this application.

Table 14.1 Classes representing manifestation or absence of diseases

Patients	Diseases						
	D_1	D_2	D_3	D_4	D_5	(...)	D_{21}
1	1	0	0	0	0	(...)	1
2	1	0	0	1	0	(...)	0
3	0	0	0	0	0	(...)	0
4	1	0	1	0	0	(...)	0
5	1	0	0	0	0	(...)	0
6	0	0	1	0	0	(...)	0
7	1	0	0	1	0	(...)	0
(...)	1	0	0	1	0	(...)	0
50	1	0	0	1	0	(...)	0

Fig. 14.2 The ART network topology used in disease diagnostics

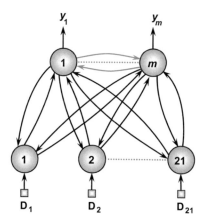

After the adjustment of its internal parameters and the identification of the m classes of treatment, the response of the ART network should indicate in which class of treatment the patient could be inserted, so that the patient could be routed to the proper treatment.

14.3 Computational Results

A set of 50 patients with disorders classified by the ICD 10, as indicated in Table 14.1, was presented to the inputs of the ART network so that it could identify the number different classes of treatment. Four simulations were made in order to compare results with respect to changes in the network vigilance parameter.

After each simulation, the ART network indicated how many classes were activated and which patient situations were in their groupings. In Fig. 14.3 the results for different vigilance parameters can be seen.

Figure 14.4 shows the number of classes of treatment required by the patients assessed by the ART network, as the vigilance parameter changes

It is possible to observe that the network is suitable for this application, since, for a set of input values, typically binary ones, the network tries to classify it into a previously defined category.

This type of application helps to define what treatments to use, as well as to route patients for a more likely diagnosis.

In Fig. 14.4 it is also possible to verify an increase in the number of classes found when the vigilance parameter increases. For a vigilance parameter around 0.5, a trend of bigger class groupings is observed, and the classes are stable between the numbers 20 and 30.

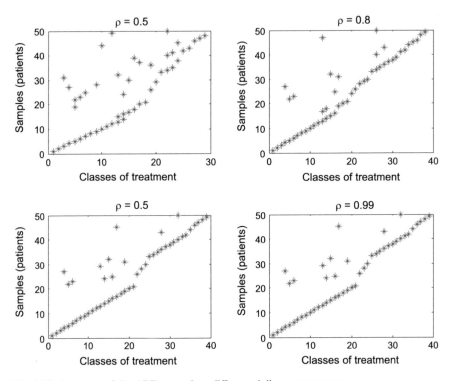

Fig. 14.3 Response of the ART network to different vigilance parameters

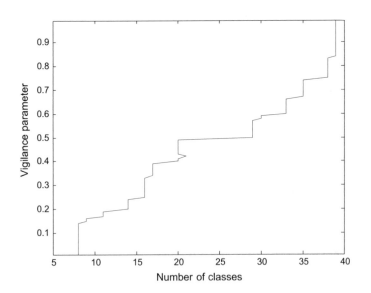

Fig. 14.4 Number of classes depending on the vigilance parameter

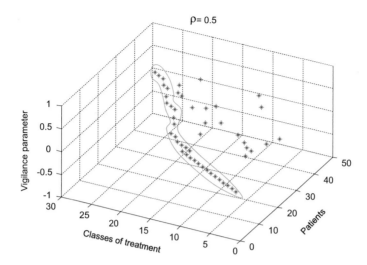

Fig. 14.5 Treatment classes for a 0.5 vigilance parameter

The grouping of patients in classes can be better examined in Fig. 14.5, in which, a similarity region between patients is illustrated, and, therefore, these patients could be routed to the same treatment groups.

The ART network showed very satisfactory results. In all cases, the network identified the number of necessary classes of treatment for the evaluated patients, contributing then to diagnose accurately such diseases.

Chapter 15
Pattern Identification of Adulterants in Coffee Powder Using Kohonen Self-organizing Map

15.1 Introduction

The purpose of this application is to identify adulterant patterns found in samples of roasted ground coffee using the Kohonen network. The network results were compared to experimental results from the Ali-C equipment (Fig. 15.1).

The use of artificial neural networks can facilitate the detection of frauds in coffee powder, which can reach up the level of 85 %. The law allows only 1 % of impurities ("peels and sticks").

To identify patterns of adulterants, a Kohonen self-organizing map (SOM) was implemented. The dataset was obtained from the Food and Coffee analyzer (Ali-C) developed by EMBRAPA (Agricultural Instrumentation).

This device identifies the presence of impurities by the photothermal principle. White light is applied to the sample surface with a frequency or pulse modulation. A heat wave is generated, and it passes through the sample until it gets to a pyroelectric detector. The electrical signals obtained depend on the structure (compression) and composition of the coffee powder. Based on this relationship, it is then possible to classify the samples.

The traditional coffee classification process regarding impurities is a craft process, this is, the samples are subjected to a previous treatment with chemical agents and, then, a visual inspection with magnifying glasses is made.

Therefore, an automatic method to perform this identification allows a better quality control without compromising the sample.

© Springer International Publishing Switzerland 2017
I.N. da Silva et al., *Artificial Neural Networks*,
DOI 10.1007/978-3-319-43162-8_15

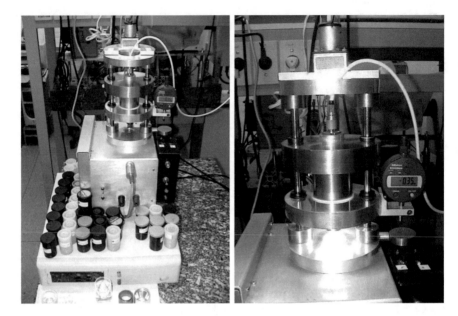

Fig. 15.1 Ali-C equipment for identifying adulterant patterns

15.2 Characteristics of the Kohonen Network

Data used in the network are physical quantities, such as electric voltage and compression degree. The Kohonen map was used for identifying adulterants in coffee samples with impurities, such as straw and grounds, and its adjustments was made as follows:

- **Input layer** → the input values were normalized to the range [0, 1] for conditioning purposes. The following descriptors were considered:

  – **Voltage** → Defined in a range of 3.0–5.0 mV. Values close to 3.0 mV indicates a high level of adulteration in the coffee powder and levels close to 5.0 mV indicates a pure coffee.
  – **Compression** → Compression ranges from 2.5 to 4.0 mm, according to the grain size of the sample.
  – **Impurity Level** → The powder is considered a pure coffee when it has known origins and did not have any trace of adulterants. It is measured with the Ali-C equipment, whose lower limit for detection is 5 %. Impurity level can range from 0 to 100 %. This parameter is used as a reference for the measurements.

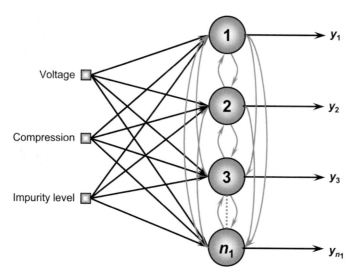

Fig. 15.2 Kohonen neural network topology used in the classification process

- **Output Layer** → Represents the winner neuron indicating one of the following classes:

  - *Class A* → Product with no adulteration.
  - *Class B* → Product adulterated with straw.
  - *Class C* → Product adulterated with coffee grounds.

The neural network topology can be best understood in Fig. 15.2.

The Euclidian norm was used as a metric to evaluate the similarity between patterns in the self-organizing phase of the Kohonen map. Two-dimensional topologies were tested with 4×4, 10×10, and 30×30 neurons. Figure 15.3 illustrates the two-dimensional topological map configuration with 4×4 neurons ($n_1 = 16$).

The topology that presented most promising results was the one with 30×30 neurons, whose final map is illustrated in Fig. 15.4, and exhibits a clear distinction between the three clusters (A, B, and C), which corresponds to the real classes.

A dataset of 90 samples was used for the adjustment phase, with a learning rate of 0.01. A propagation radius equal to three neighbors was used to define the winner neuron influence. For the test phase, a total of 10 samples were used, which were not part of the training set.

Fig. 15.3 Topological map
grid with 4 × 4 neurons

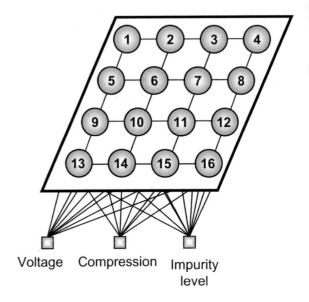

Fig. 15.4 Grouping of
classes in the Kohonen Map

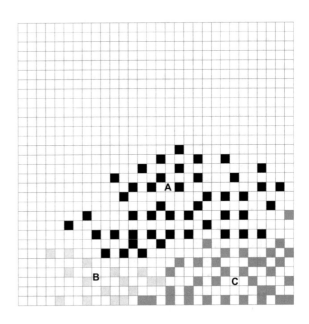

Table 15.1 Results comparison

Sample	Amplitude mean (voltage signals)	Correction for thickness (Compression)	Purity level	Classes (ANN)	Classes (Real/Ali-C)
1	0.444	0.814	1.000	A	A
2	0.678	0.465	0.656	C	C
3	0.554	0.816	0.562	B	B
4	0.885	0.565	1.000	A	A
5	0.728	0.470	0.684	C	C
6	0.551	0.696	0.555	B	B
7	0.674	0.919	1.000	A	A
8	0.610	0.996	1.000	A	A
9	0.631	0.596	0.686	B	B
10	0.786	0.513	0.712	C	C

15.3 Computational Results

The resulting clusters are shown in Table 15.1.

By visual inspection of the validation results from the previous table, it can be noticed that the map correctly groups 100 % of the samples. It must be stressed that the three clusters identified by the network are consistent with the real classes.

In summary, the results obtained by the Kohonen network were considered very satisfactory for the correct separation of the samples into classes, since it corroborates with the results from Ali-C.

Patterns related to coffees without adulterants are concentrated close to a single neuron in cluster A (Fig. 15.4). It can be noted that this cluster border is visually well defined with respect to clusters B and C, which identify the adulterated coffee. Groupings of adulterated coffee were different but very close, almost resulting in a single cluster.

Chapter 16
Recognition of Disturbances Related to Electric Power Quality Using MLP Networks

16.1 Introduction

The technical and scientific evaluation of power quality (PQ) had a great advance after the use of intelligent systems. Investigation of several abnormalities related to PQ, using such systems, allowed a much broader identification of these phenomena, enabling the systems to distinguish transitory, short-term instabilities, long-term instabilities, waveform disturbances, voltage fluctuations, voltage imbalance, and frequency changes.

Lack of power quality can lead to failure or bad operation of consumer's equipment, causing huge financial losses. Concerned with this issue, competent agencies from the Brazilian Electrical Sector created some resolutions to evaluate the quality of the supplied power. The correct identification of the disturbances that degrades power quality also allows the identification of their causes. With the above, quality indicators can be improved when properly identified, and consequently, solved.

Disturbance classification requires the extraction of specific features from electric signals, in order to identify to which class the disturbance belongs to, what is a very complex task. The extraction of features by conventional techniques, such as the Fourier transform, calculations of the RMS (Root Mean Square) value, and others, do not offer good results. Therefore, the use of artificial neural networks for this purpose becomes justifiable, since one of their main applications is in pattern recognition, and, moreover, artificial neural networks effectively approach several types of nonlinear problems.

In this chapter, it is presented an application of artificial neural networks for the recognition of four classes of disturbances related to PQ on voltage signals.

A disturbance is characterized as a change in the voltage or current that makes the signal waveform different from a sinusoidal waveform with just a fundamental frequency (first harmonic, usually, 50 and 60 Hz). The disturbances analyzed by the neural approach are described next.

© Springer International Publishing Switzerland 2017
I.N. da Silva et al., *Artificial Neural Networks*,
DOI 10.1007/978-3-319-43162-8_16

Fig. 16.1 Example of
voltage sag

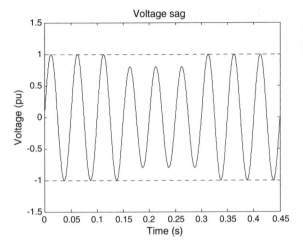

Fig. 16.2 Example of a
voltage swell

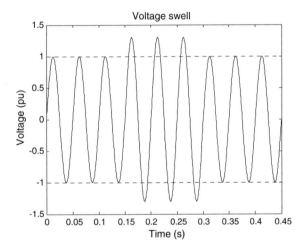

(A) **Voltage Sag**

Any decrease in the voltage magnitude that results in voltage levels between 0.1 and 0.9 pu. It arises from an electrical system fault (short circuit). Figure 16.1 illustrates an example of voltage sag.

(B) **Voltage Swell**

Any increase in the voltage magnitude that results in voltage levels between 1.1 and 1.8 pu. It can also be caused by the occurrence of a fault. Figure 16.2 shows an example of voltage swell.

(C) **Voltage Interruptions**

The interruption condition is characterized by voltage levels below 0.1 pu, and usually occurs when there is a permanent fault in the electric system. Figure 16.3 shows an example of voltage interruption.

Fig. 16.3 Example of a
voltage interruption

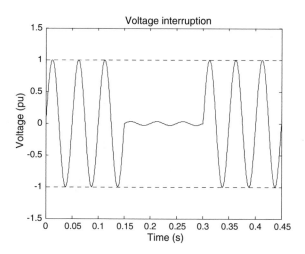

(D) **Harmonic Distortion**

Harmonic distortion occurs when the voltage signal is composed of other frequency spectrums besides the fundamental component. It is caused by transient on the electric network or nonlinear loads. Figure 16.4 shows an example of a harmonic distortion.

From this last figure, it can be noticed that the first three cycles of the signal represent a voltage signal in normal operating condition, being the subsequent three cycles an example of a harmonic distortion disturbance.

16.2 Characteristics of the MLP Network

The neural network used in this application has the purpose of recognizing the four types of disorders related to power quality, as presented in the previous section. An MLP network was used to classify these disorders from samples of the voltage signal. Therefore, the MLP topology is as follows:

- **Input Layer** → 128 inputs, corresponding to the sampling of half a cycle of the voltage signal.
- **Hidden Layer** → 5, 10, 15, or 20 neurons.
- **Output Layer** → 4 neurons, representing the 4 classes corresponding to each one of the disturbances analyzed.

The neural network topology can be better understood in Fig. 16.5.

The parameter n_1 refers to the number of neurons in the hidden layer and may assume the values 5, 10, 15, and 20.

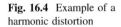

Fig. 16.4 Example of a
harmonic distortion

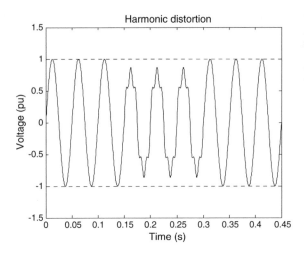

16.3   Computational Results

Through the ATP program (Alternative Transients Program), a total of 98 cases of disorders related to PQ were generated, regarding an electric distribution system. Abnormal situations were divided into 23 cases of sag, 29 cases of swell, 21 cases of interruption, and 25 cases of harmonic distortion. Then, this total (98 cases) was divided into 10 sets so as to evaluate the training and testing phases of the MLP.

All types of disturbance phenomena were used to form each one of the 10 sets. For each disturbance phenomena, 10 % of the cases were used to form the test sets, and 90 % of the cases to form the training sets.

The MLP with the best training performance was chosen and used in the validation test, whose results are presented below. Four topologies of MLP were investigated, being held 10 training processes for each one of them.

All tested topologies exhibited a hit rate of 100 %, because, by adopting four neurons in the output layer, as outlined in Sect. 5.4.1, a greater ability to delimit boundary regions is achieved. In this circumstance, there is an active neuron for each class, which improves the network efficiency. Consequently, promising results are obtained in the recognition of phenomenas related to power quality.

In summary, it is noteworthy that the use of 128 samples (half cycle of voltage signal) as inputs to the MLP requires a complex topology. Another strategy is to use preprocessing techniques to reduce the number of network inputs. Figure 16.6 presents a simplified scheme of a preprocessing strategy for the recognition of disturbances through an MLP.

It must be stressed that these investigations involving hybrid architectures, combining signal processing techniques and neural networks, have lately stood out in applications regarding the recognition of power quality phenomena.

Fig. 16.5 Topology of the neural network

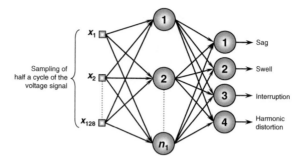

Fig. 16.6 Alternative topology for investigating PQ disturbances

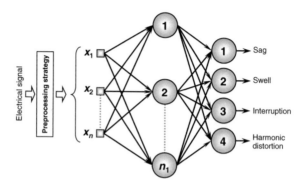

Fig. 16.7 Expert artificial neural networks for disturbance classification

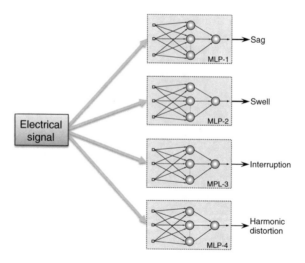

Additionally, other strategies can improve the performance of the network and reduce the network complexity in applications involving the classification of PQ problems.

Among these strategies, the use of expert networks for each one of the disturbance class can be pointed out, as shown in Fig. 16.7. With such a scheme, it would also be possible to recognize more than one disturbance at a time in the electric signal. Therefore, each expert network is responsible for recognizing some particular characteristics of each disturbance, and results produced are very promising for analyzing the occurrence of simultaneous disturbances.

In comparative terms, this configuration shown in Fig. 16.7 provides better results than other strategies, as well as, reduces the computational effort necessary for the training phase of the networks and for the adjustments made in the neural topologies.

Chapter 17
Trajectory Control of Mobile Robot Using Fuzzy Systems and MLP Networks

17.1 Introduction

Nowadays, several control methods are used in automation; however, not all of them respond in an expected way if the plant does not present a linear response.

The purpose of this application is to investigate the model, analyze the results and verify the efficiency of a controller, based on an MLP neural network, for mapping an autonomous navigation. The network training is performed using a high number of samples generated from a fuzzy inference system, which was especially projected to this function. The basic aspects related to fuzzy inference systems, together with their capabilities, can be found in Pedrycz and Gomide (2007), Buckley and Siler (2004), Ross (2010).

The fuzzy inference system was chosen to generate training patterns for the network so that the latter could absorb the decision-making characteristics. These types of problem have a high complexity, and, a mathematical modeling with an analog or digital implementation of a conventional PID controller is not always effective. The robot that received the neural controller is the Mecatron II, presented in Fig. 17.1, developed specifically to experimental essays involving optimization and control of trajectories.

The Mecatron II has light sensors and distance sensors. The light sensors were strategically positioned to receive direct light radiation from all directions, in this case, defined as Right, Left, Front and Rear. The robot must follow through a limited environment, made up by objects placed in several different positions, and by a light source.

The light source is strategically positioned in a way that the sensor radiation intake is made possible, as illustrated in Fig. 17.2.

The robot must find the light source regardless of its initial position. As an example, if the robot is positioned in the opposite direction of the light source, the rear light sensor will inform this to the neural controller, which will then make the robot turn and follow the source.

© Springer International Publishing Switzerland 2017
I.N. da Silva et al., *Artificial Neural Networks*,
DOI 10.1007/978-3-319-43162-8_17

Fig. 17.1 The Mecatron II
robot

Fig. 17.2 Mobile robot
navigation environment

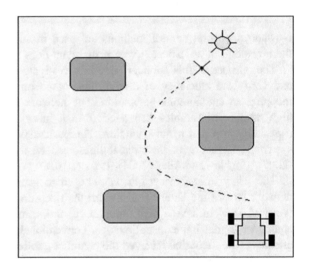

When the robot gets very close to the light source, a sound signal is emitted to indicate the conclusion of the navigation.

Therefore, the fuzzy controller has the following navigation characteristics:

(a) Position control (navigation direction).
(b) Speed control (duty cycle adjustment of the PWM (Pulse-Width Modulation controller).
(c) Arrival indicator (sound signal emission).

Because of hardware limitations, the neural controller is implemented in a support microcomputer, which has the purpose of processing the data from the Mecatron II, and then compiles and retransmits signals for trajectory control to the robot by a serial port communication.

17.2   Characteristics of the MLP Network

The neural controller is fed with information from the fuzzy inference system, since an extensive database was generated from trajectory control situations that were previously handled by the fuzzy system.

If an obstacle is not detected near the robot and the light intensity in the front sensor is low, then the robot moving speed will be high, which is controlled according to the relative distance between an obstacle and the robot, or, according to the light intensity received by the sensor.

When closer to an obstacle, the choice of which direction to take, left or right, is made according to the intensity of light reaching the corresponding side sensors. On the other hand, when the light intensity in the front sensor is high, then there is the indication that the robot is near the light source and, as a consequence, the speed will tend to zero, and a sound signal will be emitted to indicate the trajectory completion.

As all inputs and outputs are normalized, the obtained results are always in the range [0; 1]. Hence, for the output *Direction,* it was defined that values higher than 0.7 are considered as *Right,* values between 0.7 and 0.3 as *Center* and values smaller than 0.3 as *Left.*

The sound signal is emitted when the output *buzzer* assumes a value bigger than 0.7 or when the output *speed control* is directly applied to the duty cycle control of the PWM signal, which is responsible for controlling the robot motor. The neural network topology used as the controller for the Mecatron II is presented as follows:

- **Input Layer** → 5 inputs, 4 of them are from the light sensor data and 1 from the *distance sensor* (sonar) data.
- **Hidden Layer** → 45 neurons.
- **Output Layer** → 3 neurons, representing the *direction, speed control,* and *buzzer.*

The MLP network used for controlling the Mecatron II can be better understood in Fig. 17.3.

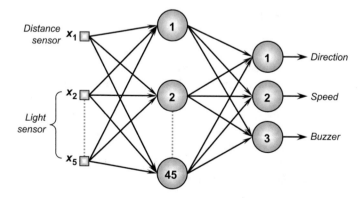

Fig. 17.3 Topology of the MLP for controlling the Mecatron II robot

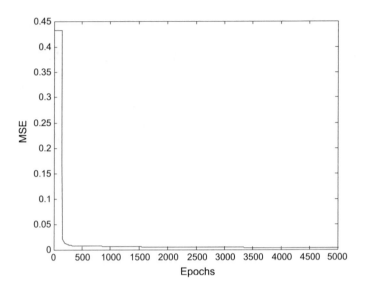

Fig. 17.4 Mean squared error behavior with respect to the number of training epochs

Data from the fuzzy inference system are considered ideal, therefore, in practice, very similar results should be obtained with artificial neural networks. For this purpose, 700 patterns containing simulation responses of the Mecatron II sensors were adopted.

In order to evaluate the performance of the neural learning process, the Mean Squared Error (MSE) behavior with respect to the training epochs can be seen in Fig. 17.4.

It can be noticed that the learning process of the MLP network is very fast, since it starts stabilizing around 250 training epochs.

17.3   Computational Results

After the MLP is trained, a set of 50 samples of values acquired from the sensors is presented to both controllers (fuzzy and neural). Figures 17.5, 17.6, and 17.7 depict the responses obtained by the neural controller and the expected responses (obtained by the fuzzy controller).

By analyzing these three graphs, it is possible to verify the quality of the control provided by the MLP controller with respect to the standard fuzzy controller.

To monitor the performances of the controllers, a graphical interface was also developed to track, online, all parameters from the neural controller, including its graphical responses obtained in real time.

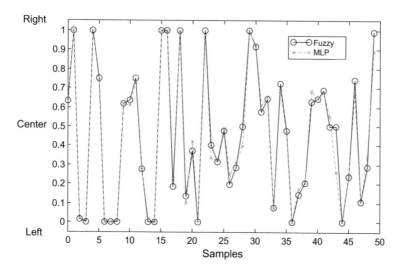

Fig. 17.5 Direction control from the neural and fuzzy controllers

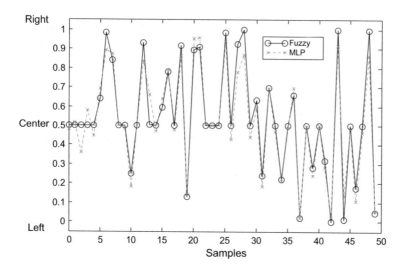

Fig. 17.6 Speed control from the neural and fuzzy controllers

The hardware is connected to a serial bus (RS232) and communicates with the software with an 115,200 bps speed. That is enough for exchanging information for the control without loss of information from the sensors.

Accordingly, it is then possible to monitor, in real time, the decisions from the neural controller, which receives information from the sensors of the Mecatron II and then sends the control signals back to the robot.

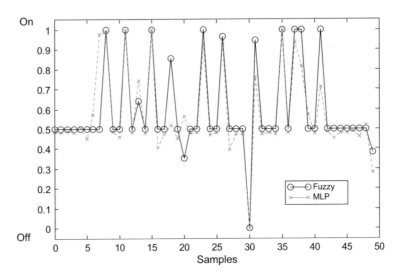

Fig. 17.7 Buzzer control from the neural and fuzzy controllers

It must be stressed that the neural control system was implemented in a microcomputer rather than in the Mecatron II microcontroller. Thus, the communication system between both platforms handles the sensor data to the neural controller, while the control signals are handled back from the microcontroller to the robot.

In summary, through this computational analysis, it can be observed the efficiency of the intelligent controller based on MLP networks, which was capable of learning and absorbing characteristics from a fuzzy controller, producing efficient responses to the navigation system, whose behavior between variables is nonlinear.

Finally, the only requirement to use the neural controller is the availability of a database that can feed its learning process. For this application, such requirement has been supplied by simulations involving the sensors behaviors through a fuzzy inference system.

Chapter 18
Method for Classifying Tomatoes Using Computer Vision and MLP Networks

18.1 Introduction

Quality control in a production line, when performed in a nonautomatic way, is subjected to mistakes made by the inspection agents, who over time and due to fatigue let products that are improper for human consumption pass through the mats.

The majority of automatic inspection systems is based on human vision and has a simulated artificial vision to perform product classification. Inspection systems with computer vision are used mainly on agricultural applications, which have a high correlation between food quality and its appearance.

Automatic quality inspection of tomatoes has as its main motivation the dissatisfaction of consumers, as reported in several technical reports from the food industry. Therefore, the use of artificial neural networks for classifying tomatoes can make the inspection process more efficient, improving the quality of the products that reach consumers.

The 102 tomato samples used for this application were acquired in supermarkets from the São Carlos region (Brazil). They belong to the "Saladete" group, whose standard pattern is rounded and a final reddish color. Figure 18.1 illustrates some tomatoes from this group.

The tomato samples obtained had different colors and sizes and were grouped into four classes as shown in Table 18.1.

Images from the tomatoes were done by a digital camera, using an acquisition module that also has a diffuse lightning camera, whose purpose is to reduce, maximally, unwanted reflections. The acquisition module can be seen in Fig. 18.2.

In Fig. 18.3, an example of a digital image obtained by the module above is shown.

© Springer International Publishing Switzerland 2017
I.N. da Silva et al., *Artificial Neural Networks*,
DOI 10.1007/978-3-319-43162-8_18

Fig. 18.1 Tomatoes from the "saladete" group in different maturation phases

Table 18.1 Distribution of tomato classes

Classes	Equatorial diameter (cm)	Color
A	Bigger than 7	Completely red
B	Smaller than 7	Completely red
C	Bigger than 7	Greenish spots
D	Smaller than 7	Greenish spots

18.2 Characteristics of the Neural Network

From the digital images, it is possible to extract features that will be presented to the neural network for the learning process. From this process, six normalized variables were obtained, which represent the average pixels related to the colors Red (R), Green (G), and Blue (B), as well as the chromaticity coordinates of red (r), green (g), and blue (b). The chromaticity coordinates (r, g, b) are obtained from expression (18.1), while the average pixels (R, G, B)$_m$ are calculated as (18.2), that is:

$$(r, g, b) = \frac{\sum_{i=1}^{N} (R, G, B)_i}{\sum_{i=1}^{N} R_i + \sum_{i=1}^{N} G_i + \sum_{i=1}^{N} B_i} \tag{18.1}$$

Fig. 18.2 Lightning and
image acquisition module

Fig. 18.3 Example of an
image acquired by the module

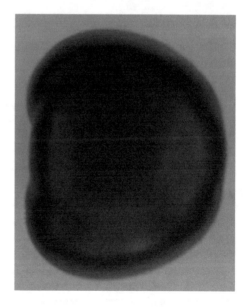

$$(R, G, B)_m = \frac{\sum_{i=1}^{N} (R, G, B)_i}{N \times 255},$$ (18.2)

where N is the number of pixels.

By running through the indexes from expressions (18.1) and (18.2), all average pixels and chromaticity coordinates are calculated. The process for feature extraction from the digital images, which the basic concepts are found in Parker (2010), is summarized as the following algorithm:

1. Colored image acquisition.
2. Application of a average filter with three neighboring pixels.
3. Conversion of the color image to a grayscale image.
4. Obtaining segmentation threshold via Otsu (1979) algorithm.
5. Image binarization.
6. Complementing processing to obtain a black background and a white object.
7. Multiplication of the original colored image by the binary image in order to obtain a segmented colored image.
8. Separation of each RGB channel into three vectors, so that they contain only pixels from the object.
9. Computation of average pixels, followed by normalization, as expression (18.2).
10. Obtaining the chromaticity coordinates, as expression (18.1).

The MLP network used for this phase has the following topological characteristics:

- **Input Layer** → seven inputs, being six of them color attributes (three of average pixels and three of chromaticity coordinates) and one input related to the equatorial diameter of the tomatoes.
- **Hidden Layer** → 15 neurons.
- **Output Layer** → two neurons, whose combined outputs are used to produce the tomato classification, corresponding to the four classes (A, B, C, or D) defined in Table 18.1.

This topology with a single hidden layer can be better understood in Fig. 18.4, where the color attributes are given by the six components calculated by expressions (18.1) and (18.2), this is, three components related to the chromaticity coordinates and three to the average pixel calculation, respectively.

Although this first configuration, presented in Fig. 18.4, can be used to perform the tomato classification, a second configuration was tested, which is denominated "cascaded", and whose purpose is to verify whether there is a substantial improvement in hit rates.

This second configuration is composed of two cascaded MLP networks, this is, the responses produced by the first MLP network are the inputs to the second network. The first network has its topology represented as follows:

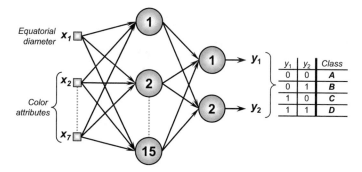

Fig. 18.4 Topology of the MLP for tomato classification

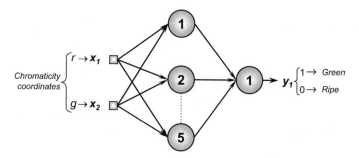

Fig. 18.5 Topology of the first MLP to classify tomatoes between Green and Ripe (cascaded configuration)

- **Input Layer** → two inputs, being them the chromaticity coordinates $\{r\}$ and $\{g\}$, since they are the features of interest because of the predominance of the colors *Red* and *Green* in the classes *Ripe* and *Green*, respectively.
- **Hidden Layer** → five neurons.
- **Output Layer** → one neuron, whose output will produce the unitary value $\{1\}$ to identify the *Green* class, whereas it will produce a null value $\{0\}$ to indicate the *Ripe* class.

The first MLP network that comprises the second tested configuration (cascaded) is shown in Fig. 18.5.

Once the tomato samples are classified in *Green* and *Ripe*, the variable *Equatorial diameter* can then be considered to classify the tomatoes into the four varieties described above. The second MLP network that comprises the cascaded configuration, as illustrated in Fig. 18.6, has the following topology:

- **Input Layer** → two inputs, being one of them the answer $\{y_1\}$ produced by the first MLP network (Fig. 18.5); the other input is the *Equatorial diameter* measurement of the tomato samples.
- **Hidden Neural Layer** → five neurons.

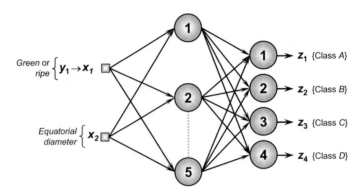

Fig. 18.6 Topology of the second MLP to classify tomatoes (cascaded configuration)

- **Output Neural Layer** → four neurons, whose outputs are used for classifying the tomatoes into one of the four previously defined classes.

Regarding the learning process, all artificial neural networks in this application used 90 % of the available samples for the training phase, whereas the remaining samples {10 %} were used for the testing phase.

The chosen approach for the answers produced by the output neuron of the second MLP network (Fig. 18.6) is the "one of c-classes" strategy, as described in Sect. 5.4.1.

18.3 Computational Results

The first MLP configuration tested presented hit rates of 80 %. The number of neurons was also increased to enhance results, achieving a final efficiency of 90 %. However, from this value of performance on, its topological structure started to become complex, being incapable of increasing the hit rates already obtained.

As to the second configuration (two cascaded MLP networks), the classification between *Green* and *Ripe*, done by the first MLP network presented a hit rate of 100 % during the test phase. This rate was also achieved by the second MLP, which was responsible for classifying the tomatoes into the four varieties. Hence, the solution option with cascaded neural networks was more efficient.

In summary, one of the main tasks associated with this application is the process of selecting the best variables to be presented to the networks, so that they can perform the correct tomato classification. The association of two cascaded networks, to split the classification task, proved to be more computationally efficient, obtaining better results when compared to using a single neural network.

Chapter 19
Performance Analysis of RBF and MLP Networks in Pattern Classification

19.1 Introduction

Both the RBF network and the MLP network can be used as pattern classifiers in situations where there are high-dimensional input spaces.

Analyzes of what network type would produce better results, as well as, what topology would be more appropriate to achieve such performances, are of great value to choose which network is the most suitable for each type of problem.

Pattern recognition done by human beings is derived from several correlations made by sensory organs. Of course, human begins are excellent pattern classifiers, doing such tasks with minimal effort. Similarly, using a learning process, artificial neural networks can assign classes by processing input signals.

So, when a new sample is presented, the networks can extract necessary information to perform the classification, since they have a generalization ability, which was acquired through that process of learning from examples.

MLP networks, presented in detail in Chap. 5, have one or more intermediate layers, where each neuron is under the action of a nonlinear activation function. The high connectivity level between these neurons is also an inherent characteristic of this neural architecture.

With regards to RBF networks, as exposed in details in Chap. 6, they typically have a single intermediate layer, where each neuron uses the radial basis activation function. The output layer, in turn, uses the linear activation function to produce the network answer from the input signals.

Typically, hidden layer neurons of an MLP network follow the same model. Hidden layer neurons of an RBF network follow different models and could aim different purposes.

© Springer International Publishing Switzerland 2017
I.N. da Silva et al., *Artificial Neural Networks*,
DOI 10.1007/978-3-319-43162-8_19

Table 19.1 Distribution of samples into classes (Wine database)

Classes	Frequency (%)
1	33.14
2	39.88
3	26.27

Table 19.2 Distribution of samples into classes (Wisconsin database)

Classes	Frequency (%)
Benign	65.0
Malign	35.0

19.2   Characteristics of the MLP and RBF Networks Under Analysis

The input signals used to accomplish the comparison of performances, between the MLP and RBF networks, are derived from two public and noticeable databases, i.e., the Wine and Breast Cancer Wisconsin databases.

The Wine database is identified by 14 numeric attributes, being 13 inputs and one output, which represent three distinct wine classes. The distribution of these classes is given in Table 19.1.

Wisconsin database also has numeric attributes, from biopsy analysis in patients, being nine inputs and one output. Samples are classified as Benign and Malign, which frequencies are shown in Table 19.2.

It must be pointed out that both databases have long been used in investigations involving comparative analysis between methods for pattern classification.

19.3   Computational Results

The same number of neurons in the intermediate layer was considered for comparing performances of the different topologies of both the MLP and RBF networks. For the MLP networks only single layer topologies were tested. A total of 90 % of the available samples were used for the training phase, whereas, the 10 % remaining samples were used in the test phase.

By analyzing Fig. 19.1, it is then possible to state that, when considering the Wine database samples, the MLP networks presented a better result during the training phase with topologies up to three neurons in the intermediate layer, since the deviations in the training process were smaller for such configurations.

As illustrated in Fig. 19.2, the MLP topologies, taking into account a momentum term, were also evaluated to verify the necessary number of epochs for convergence.

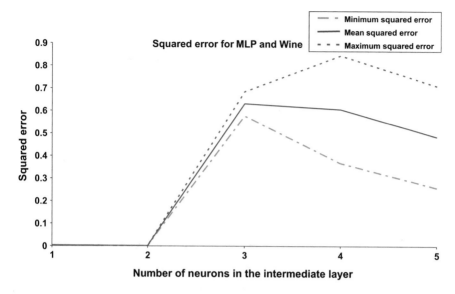

Fig. 19.1 Training results of the MLP network (Wine database)

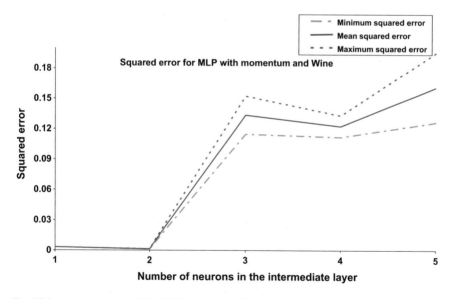

Fig. 19.2 Training results of the MLP networks with a momentum term (Wine database)

In this case, it can be observed that better training results with reduced deviation values can be seen for MLP topologies with 1, 2, and 4 neurons in their intermediate layers.

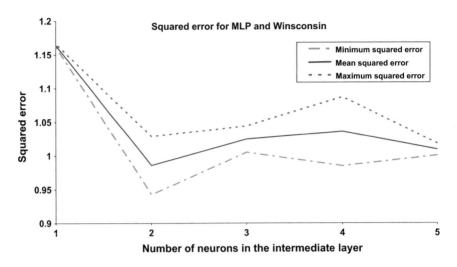

Fig. 19.3 Training results of the MLP networks (Wisconsin database)

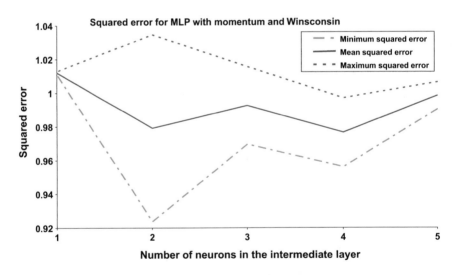

Fig. 19.4 Training results of the MLP networks with a momentum term (Wisconsin database)

In Fig. 19.3 it is shown the training results for the MLP networks, assuming now the Wisconsin database.

For the Wisconsin database, an MLP network with a momentum term was also implemented, and the results can be found in Fig. 19.4.

For the topologies presented in Fig. 19.4, considering deviation values, it can be observed that configurations with 1 and 5 neurons in the intermediate layer provide

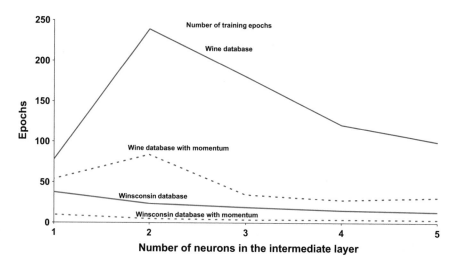

Fig. 19.5 Number of epochs necessary for training the various MLP networks

Table 19.3 Hit rates obtained in MLP test phase

Neurons in intermediate layer	Hit rates (%) Wine		Hit rates (%) Winsconsin	
	Without momentum	With momentum	Without momentum	With momentum
5	100.0	100.0	98.6	98.6
4	100.0	100.0	98.6	98.6
3	100.0	100.0	98.6	98.6
2	83.3	81.5	98.6	98.6
1	53.3	44.4	71.0	71.0

the best results. In Fig. 19.5 it is possible to identify the average number of epochs required to accomplish each one of the MLP training processes.

For both investigated databases, it can be noted a significant reduction of the number of training epochs when a momentum term is inserted. Hit rates obtained by the MLP networks can be seen in Table 19.3.

Based on these results, it can be stated that, when considering the Wine database, the best MLP topology features three neurons in its intermediate layer and uses the momentum term in its training process. In regards to the Wisconsin database, the momentum term must also be used to speed up the training process, having the best results for a topology with two neurons in the intermediate layer. These topologies were chosen by considering the network complexity as well as the efficiency obtained in the tests.

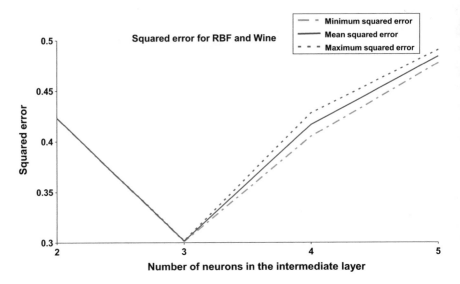

Fig. 19.6 Training results of the RBF networks (Wine database)

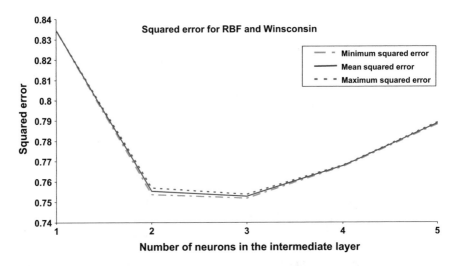

Fig. 19.7 Training results of the RBF networks (Wisconsin database)

In Figs. 19.6, 19.7, and 19.8 are shown the results obtained by RBF networks, also in relation to the same basis of Wine and Wisconsin.

Hit rates obtained during the test phases of the RBF networks can be examined in Table 19.4.

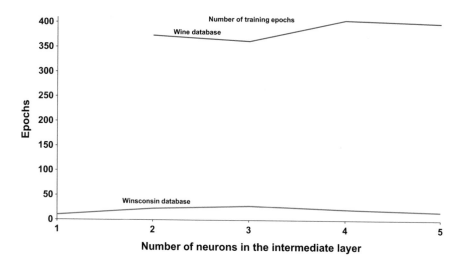

Fig. 19.8 Number of required epochs for training the several RBF networks

Table 19.4 Hit rates obtained in the RBF test phase

Number of neurons in the intermediate layer	Hit rates (%) Wine	Hit rates (%) Wisconsin
5	100.0	89.7
4	94.4	91.3
3	100.0	94.2
2	88.9	95.7
1		95.7

By analyzing all results obtained by the RBF networks, the topology with three neurons in the intermediate layer can be chosen as the best classifier for the Wine database, whereas the topology made up by only one neuron is the most suitable for the Wisconsin database.

In summary, through this brief comparative analysis, it can be seen that the listed results, regarding hit rates, were very similar between both these networks. This fact was already expected, since both MLP networks and RBF networks are universal curve fitting tools, which allows an RBF to operate in a similar way to a particular MLP.

However, very different behaviors can be observed between the networks in the training process, especially with the momentum term inclusion in the MLP topologies, resulting in a considerable acceleration of their convergences.

On the other hand, the RBF networks required fewer neurons than the MLP networks, when applied in the Wisconsin database classification.

Chapter 20
Solution of Constrained Optimization Problems Using Hopfield Networks

20.1 Introduction

Constrained optimization problems refer, generally, to maximize or minimize an objective function subjected to a set of equality, and/or inequality constraints (linear or nonlinear) (Bazaraa and Shetty 1979).

Although there are several mathematical programming methods in the literature that can be applied for solving constrained optimization problems, there is a growing need to investigate alternative methods that exploit parallel and adaptive processing architectures. Accordingly, artificial neural networks become a promising approach that can be efficiently applied to this type of problems. Among the main advantages of using the artificial neural network approach to constrained nonlinear optimization, the following stands out: (i) intrinsic ability of parallel operation; (ii) easy hardware implementation; (iii) achievement of high computation rates by simple processing elements.

The problem addressed in this chapter consists of applying a Hopfield network to minimize an energy function E^{op}, which represents the problem objective function, subjected to equality/inequality constraints. These restrictions are all grouped in a single energy term, named E^{conf}, which has the purpose of confining all constraints involved with the problem. Therefore, Hopfield network equilibrium points represent possible solutions to this optimization problem.

As seen in Chap. 7, the expression that dictates the continuous-time behavior of each neuron in the Hopfield network is given by:

$$\dot{u}_j(t) = -\eta \cdot u_j(t) + \sum_{i=1}^{n} W_{ji} \cdot v_i(t) + i_j^b, \text{ with } j = 1, \ldots, n \qquad (20.1)$$

$$v_j(t) = g(u_j(t)), \qquad (20.2)$$

© Springer International Publishing Switzerland 2017
I.N. da Silva et al., *Artificial Neural Networks*,
DOI 10.1007/978-3-319-43162-8_20

where:

$\dot{u}_j(t)$ is the internal state of the jth neuron, being $\dot{u}(t) = du/dt$;

$v_j(t)$ is the output of the jth neuron;

W_{ji} is the value of the synaptic weight connecting the jth neuron to the ith neuron;

i_j^b is the bias applied to the jth neuron;

$g(.)$ is an activation function, monotonically increasing, that restricts each neuron output in an interval defined previously;

$\eta \cdot u_j(t)$ is the passive decay term.

The network equilibrium points, as presented in Sect. 7.3, correspond to the $v(t)$ values that minimize the following energy function associated with the Hopfield network, that is:

$$E(t) = -\frac{1}{2}v(t)^T \cdot W \cdot v(t) - v(t)^T \cdot i^b \tag{20.3}$$

Therefore, the solution of optimization problems when using a Hopfield network consists of determining, for each type of problem, the weight matrix W and the bias vector i^b associated to the network energy function (20.3).

For constrained optimization problems, it is used an energy function comprised of two terms, as follows:

$$E(t) = E^{\text{op}}(t) + E^{\text{conf}}(t), \tag{20.4}$$

where:

$$E^{\text{op}}(t) = -\frac{1}{2}v(t)^T \cdot W^{\text{op}} \cdot v(t) - v(t)^T \cdot i^{\text{op}} \tag{20.5}$$

$$E^{\text{conf}}(t) = -\frac{1}{2}v(t)^T \cdot W^{\text{conf}} \cdot v(t) - v(t)^T \cdot i^{\text{conf}} \tag{20.6}$$

Hence, the Hopfield network has the purpose of simultaneously minimizing an energy function $\{E^{\text{op}}(t)\}$, as well as minimizing the other energy function $\{E^{\text{conf}}(t)\}$ that maps all the structural constraints of the problem.

20.2 Characteristics of the Hopfield Network

The main features of a Hopfield network for solving constrained optimization problems, which has two energy terms, are illustrated in Fig. 20.1.

As it can be seen in this figure, the dynamics of the network operation can be explained through three main stages, namely

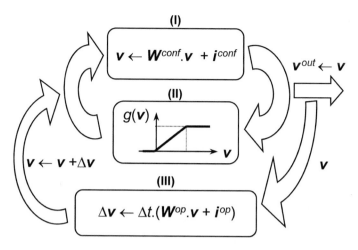

Fig. 20.1 Hopfield network for solving constrained optimization problems

(I) **Minimization of E^{conf}** → Corresponds to the projection of v onto a valid subspace that confines all constraints imposed by the problem:

$$v(t+1) = W^{\text{conf}} \cdot v(t) + i^{\text{conf}} \Rightarrow v \leftarrow W^{\text{conf}} \cdot v + i^{\text{conf}}, \qquad (20.7)$$

where W^{conf} is a projection matrix $\left(W^{\text{conf}} \cdot W^{\text{conf}} = W^{\text{conf}}\right)$ and i^{conf} is orthogonal to the valid subspace $\left(W^{\text{conf}} \cdot i^{\text{conf}} = 0\right)$. A detailed analysis of this valid-subspace technique is presented in Silva (1997), Aiyer et al. (1990).

(II) **Application of an Activation Function** → use of the "symmetric ramp" activation function for constraining v inside a previously defined hypercube:

$$g_i(v_i) = \begin{cases} \lim_i^{\text{inf}}, & \text{if } v_i < \lim_i^{\text{inf}} \\ v_i, & \text{if } \lim_i^{\text{inf}} \leq v_i \leq \lim_i^{\text{sup}}, \\ \lim_i^{\text{sup}}, & \text{if } v_i > \lim_i^{\text{sup}} \end{cases} \qquad (20.8)$$

where $v_i(t) \in [\lim_i^{\text{inf}}, \lim_i^{\text{sup}}]$. In this circumstance, although v is inside a particular set, the Hopfield network can also be structured to represent any constrained optimization problem. In such situations, we have $\lim_i^{\text{inf}} = -\infty$ and $\lim_i^{\text{sup}} = \infty$.

(III) **Minimization of E^{op}** → Change in v toward an optimized cost solution (defined by W^{op} and i^{op}) that corresponds to the network equilibrium points, which are the solutions to the optimization problem. Using a "symmetric ramp" activation function, and given that $\eta = 0$, since changes in neurons are done synchronously, i.e., all neurons change simultaneously, then Eq. (20.1) becomes:

$$\frac{d\boldsymbol{v}(t)}{dt} = \dot{\boldsymbol{v}} = -\frac{\partial E^{op}(t)}{\partial \boldsymbol{v}} \tag{20.9}$$

$$\Delta \boldsymbol{v} = -\Delta t \cdot \nabla E^{op}(\boldsymbol{v}) = \Delta t \cdot (W^{op} \cdot \boldsymbol{v} + \boldsymbol{i}^{op}) \tag{20.10}$$

Therefore, the E^{op} minimization consists in changing $\boldsymbol{v}$ in the opposite direction of the gradient of the E^{op} function with respect to $\boldsymbol{v}$. These results are also valid when a hyperbolic tangent function is used.

It must be mentioned that the application of steps (I) through (III) is carried out in a discrete manner. In this case, after the considerations made for step (III), the resolution of the differential equation shown in (20.9) is similar to the Euler method (Chapra and Canale 2014), and, regarding optimization, this represents a gradient descent process with fixed step equal to Δt.

In summary, after each optimization step made in (III), it becomes necessary to perform, iteratively, both subsequent steps (I) and (II) until there is the convergence of the output vector $\boldsymbol{v}$, ensuring, therefore, that all involved constraints were satisfied.

Hence, according to Fig. 20.1, each iteration has two very distinct stages. In the first stage, as described by Step (III), the $\boldsymbol{v}$ vector is changed using only the gradient of the E^{op} term. For the second Stage, after the realization of each topic provided by Step (III), the vector $\boldsymbol{v}$ is then projected onto the valid-subspace {Step (I)} confining all problem constraints, and, then, $\boldsymbol{v}$ is limited by the activation function {Step II} so that its elements are within the $[\lim_i^{inf}, \lim_i^{sup}]$ domain.

Finally, the convergence process is finished when the values $\boldsymbol{v}^{out}$ remain almost constant for two successive iterations, wherein this condition, the $\boldsymbol{v}^{out}$ value is equal to the $\boldsymbol{v}$ value.

20.3   Mapping an Optimization Problem Using a Hopfield Network

The following optimization problem is considered here, consisting in p inequality constraints and N variables, this is:

$$\text{Minimize } E^{op}(\boldsymbol{v}) = f(\boldsymbol{v}) \tag{20.11}$$

$$\text{Subject to } : E^{conf}(\boldsymbol{v}) : h_i(\boldsymbol{v}) \leq 0, \ i \in \{1, \ldots, p\} \tag{20.12}$$

$$z^{min} \leq \boldsymbol{v} \leq z^{max}, \tag{20.13}$$

where $\boldsymbol{v}, z^{min}, z^{max} \in \mathfrak{R}^N$; $f(\boldsymbol{v})$ and $h_i(\boldsymbol{v})$ are continuous and differentiable functions. Vectors z^{min} and z^{max}, respectively, define the lower and upper boundaries of each component in vector $\boldsymbol{v}$. Conditions (20.12) and (20.13) define a convex closed set in

$\mathfrak{R}^N$, in which the vector v must belong to in order to represent a valid solution to the optimization problem dictated by (20.11).

One answer to the problem can be obtained through a Hopfield network, in which the implementation of the valid subspace ensures compliance with the conditions established in (20.12). Furthermore, the initial hypercube represented by the inequality constraints (canalizing variables) in (20.13), is directly mapped by the "symmetric ramp" function given by (20.8), which is used as a neural activation function, this is, $v \in [z^{\min}, z^{\max}]$.

Parameters W^{conf} and i^{conf} that belong to the valid subspace are calculated by transforming the inequality in (20.12) in a set of equality constraints through the use of an auxiliary variable to each inequality constraint:

$$h_i(v) + \sum_{j=1}^{p} \delta_{ij} \cdot q_j = 0, \tag{20.14}$$

where $q_j \geq 0$ are the auxiliary variables, which can be considered as variables v_i, and δ_{ij} is defined by the Kronecker impulse function (Graham 1986), given by:

$$\delta_{ij} = \begin{cases} 1, & \text{if } i=j \\ 0, & \text{if } i \neq j \end{cases} \tag{20.15}$$

After this transformation, Eqs. (20.11), (20.12), and (20.13), which define the optimization problem, can be rewritten as:

$$\text{Minimize } E^{\text{op}}(v^+) = f(v^+) \tag{20.16}$$

$$\text{Subject to : } E^{\text{conf}}(v^+) : h_i(v^+) \leq 0, \, i \in \{1, \ldots, p\} \tag{20.17}$$

$$z^{\min} \leq v^+ \leq z^{\max}, \, i \in \{1, \ldots, N\} \tag{20.18}$$

$$0 \leq v^+ \leq z^{\max}, \, i \in \{N+1, \ldots, N^+\}, \tag{20.19}$$

where $N^+ = N + p$, and $v^{+T} = [v^T q^T] \in \mathfrak{R}^{N+}$ is a vector of extended variables. It must be noticed that E^{op} does not depend on the auxiliary variables q.

The projection matrix W^{conf}, regarding the valid-subspace equation, is calculated by the projection of v^+, obtained from the minimization of $E^{\text{op}}(v^+) = f(v^+)$, onto a subspace tangent to the surface bounded by the constraints given in (20.17). Thus, an equation for W^{conf} can be defined by (Luenberger and Ye 2016):

$$W^{\text{conf}} = I - \nabla h(v^+)^T \cdot (\nabla h(v^+) \cdot \nabla h(v^+)^T)^{-1} \cdot \nabla h(v^+), \tag{20.20}$$

where:

$$\nabla h(v^+) = \begin{bmatrix} \frac{\partial h_1(v^+)}{\partial v_1^+} & \frac{\partial h_1(v^+)}{\partial v_2^+} & \cdots & \frac{\partial h_1(v^+)}{\partial v_{N^+}} \\ \frac{\partial h_2(v^+)}{\partial v_1^+} & \frac{\partial h_2(v^+)}{\partial v_2^+} & \cdots & \frac{\partial h_2(v^+)}{\partial v_{N^+}} \\ \vdots & \vdots & \ddots & \\ \frac{\partial h_p(v^+)}{\partial v_1^+} & \frac{\partial h_p(v^+)}{\partial v_2^+} & & \frac{\partial h_p(v^+)}{\partial v_{N^+}} \end{bmatrix} = \begin{bmatrix} \nabla h_1(v^+)^T \\ \nabla h_2(v^+)^T \\ \vdots \\ \nabla h_p(v^+)^T \end{bmatrix} \qquad (20.21)$$

Substituting the matrix W^{conf}, given by (20.20), into the valid-subspace equation in (20.7), it is obtained:

$$v^+ \leftarrow (I - \nabla h(v^+)^T \cdot (\nabla h(v^+) \cdot \nabla h(v^+)^T)^{-1} \cdot \nabla h(v^+)) \cdot v^+ + i^{\mathrm{conf}} \quad (20.22)$$

Finally, results of the Lyapunov stability theory (Vidyasagar 2002) must be introduced in Eq. (20.22) to ensure the stability of the nonlinear system, and, consequently, enforce the network convergence to equilibrium points that represent a solution to the problem. From the Jacobian definition, when v tends to the equilibrium points it is implied that $v^e = 0$. In this case, the i^{conf} value must also be null to satisfy the equilibrium condition, i.e.:

$$v^e = v(t) = v(t + \Delta t) = 0 \qquad (20.23)$$

Therefore, Eq. (20.17) can be approximated by:

$$h(v^+) \approx h(v^e) + J \cdot (v^+ - v^e), \qquad (20.24)$$

where $J = \nabla h(v^+)$ and $h(v^+) = [h_1(v^+) \quad h_2(v^+) \quad \cdots \quad h_p(v^+)]^T$.

In the vicinity of the equilibrium point, where v^e can be considered null $\{v^e = 0\}$, the next relationship is valid:

$$\lim_{v^+ \to v^e} \frac{\|h(v^+)\|}{\|v^+\|} = 0 \qquad (20.25)$$

By introducing the results of (20.24) and (20.25) into (20.22), it is obtained:

$$v^+ \leftarrow v^+ - \nabla h(v^+)^T \cdot (\nabla h(v^+) \cdot \nabla h(v^+)^T)^{-1} \cdot h(v^+) \qquad (20.26)$$

Therefore, in (20.26), the synthesis of the valid-subspace expression to approach nonlinear equation systems is obtained. Given the above, for constrained optimization problems, the original valid-subspace expression in (20.7), which is represented by Stage (I) in Fig. 20.1, must be substituted by (20.26). Therefore, according to Fig. 20.1, a successive application of Stage (I), followed by Stage (II), makes v converge to an equilibrium point that fulfills all constraints imposed by the nonlinear optimization problem.

Parameters W^{op} and i^{op}, associated with the E^{op} energy term given by (20.5) and represented in (20.16), are defined so that the optimal solution corresponds to the E^{op} optimization. This procedure can be made by changing vector v^+ toward the opposite direction of the gradient of the E^{op}energy function. As (20.17) and (20.18) define a finite (limited) polyhedron, the objective function in (20.16) always has a minimum point. Therefore, the network equilibrium points can be calculated by assuming the following values for W^{op} and i^{op}:

$$i^{op} = -\left[\frac{\partial f(v^+)}{\partial v_1^+} \quad \frac{\partial f(v^+)}{\partial v_2^+} \quad \cdots \quad \frac{\partial f(v^+)}{\partial v_{N+}^+} \right]^T \tag{20.27}$$

$$W^{op} = 0 \tag{20.28}$$

As mentioned above, since $v^{+T} = [v^T \quad w^T]$, the vector i^{op} given in (20.27) can be then represented by:

$$i^{op} = -\left[\frac{\partial f(v^+)}{\partial v_1} \quad \frac{\partial f(v^+)}{\partial v_2} \quad \cdots \quad \frac{\partial f(v^+)}{\partial v_N} \quad \frac{\partial f(v^+)}{\partial q_1} \quad \frac{\partial f(v^+)}{\partial q_2} \quad \cdots \quad \frac{\partial f(v^+)}{\partial q_p} \right]^T \tag{20.29}$$

As the E^{op} optimization process does not depend on auxiliary variables q, the expression given in (20.29) can be then substituted by:

$$i^{op} = -\left[\underbrace{\frac{\partial f(v^+)}{\partial v_1} \quad \frac{\partial f(v^+)}{\partial v_2} \quad \cdots \quad \frac{\partial f(v^+)}{\partial v_N}}_{N-\text{components}} \quad \underbrace{0 \; 0 \; 0 \ldots 0}_{p-\text{components}} \right]^T \tag{20.30}$$

Some simulation results are presented in the next section to illustrate the performance of the proposed neural network.

20.4 Computational Results

In the first problem, as proposed in Bazaraa et al. (2006), the objective function is subjected to some inequality constraints, this is:

$$\text{Minimize} f(v) = e^{v_1} + v_1^2 + 4v_1 + 2v_2^2 - 6v_2 + 2v_3$$

$$\text{Subject to :} \quad \begin{aligned} v_1^2 + e^{v_2} + 6v_3 &\le 15 \quad (R1) \\ v_1^4 - v_2 + 5v_3 &\le 25 \quad (R2) \\ v_1^3 + v_2^2 - v_3 &\le 10 \quad (R3) \\ 0 \le v_1 &\le 4 \quad (R4) \\ 0 \le v_2 &\le 2 \quad (R5) \\ v_3 &\ge 0 \quad (R6) \end{aligned}$$

Fig. 20.2 Evolution of the network output vector (first problem)

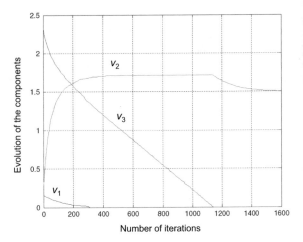

Fig. 20.3 Evolution of the objective function (first problem)

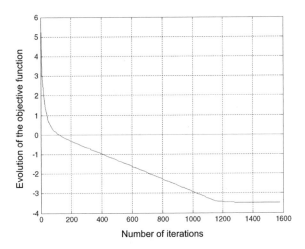

For this optimization, with three inequality constraints and with limited variables, the solution vector (equilibrium point) obtained after the Hopfield network convergence was given by $v = [0.00001\ 1.50002\ 0.00000]^T$, with the objective function $f(v) = -3.49999$. This result is very close to the optimal solution given by $v^* = [0.0\ 1.5\ 0.0]^T$ and $f(v^*) = -3.5$. Figure 20.2 illustrates the evolution of v_1, v_2, and v_3 obtained by the Hopfield network with respect to the number of iterations.

The outlined behavior of the objective function is represented in Fig. 20.3. It is worth noting that constraints given in (R4), (R5), and (R6) are directly handled by the network activation function defined in (20.8).

Fig. 20.4 Evolution of the network output vector (second problem)

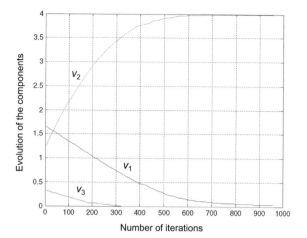

The second problem handled by the Hopfield network is made up of inequality constraints defined by the following expressions:

$$\text{Minimize} f(v) = v_1^3 + 2v_2^2 \cdot v_3 + 2v_3$$

$$\text{Subject to:} \quad v_1^2 + v_2 + v_3^2 = 4$$
$$v_1^2 - v_2 + 2v_3 \leq 2$$
$$v_1, v_2, v_3 \geq 0$$

Figure 20.4 shows the evolution of v_1, v_2, and v_3 with respect to the number of iterations. The solution vector obtained after the network convergence is given by $v = [0.00000 \ 4.00000 \ 0.00001]^T$, being the final value of the objective function $f(v) = E^{\text{op}}(v) = 0.00034$. The optimal solution for the problem is given by $v^* = [0.0 \ 4.0 \ 0.0]^T$, with $E^{\text{op}}(v^*) = 0.0$.

Figure 20.5, in turn, illustrates the behavior of the objective function with respect to the number of iterations. Initial values assigned to vector v were randomly generated between zero and one.

The last example aims to confirm the applicability of the proposed network when all constraints associated with the problem are linear, this is:

$$\text{Minimize} f(v) = 0.4v_2 + v_1^2 + v_2^2 - v_1 v_2 + \tfrac{1}{30} v_1^3$$

$$\text{Subject to:} \quad v_1 + 0.5v_2 \geq 0.4$$
$$0.5v_1 + v_2 \geq 0.5$$
$$v_1, v_2 \geq 0$$

Figure 20.6 shows the evolution of the elements in the output vector with respect to the number of iterations. After the network convergence, the solution vector obtained is given by $v = [0.3398 \ 0.3301]^T$, with $E^{\text{op}}(v) = 0.2456$. These results are

Fig. 20.5 Evolution of the
objective function (second
problem)

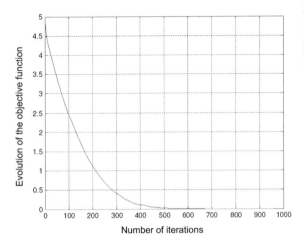

Fig. 20.6 Evolution of the
network output (third
problem)

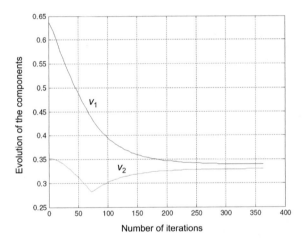

very near to the exact solution values, given by $v^* = [0.3395 \ 0.3302]^T$, with
objective function $E^{OP}(v^*) = 0.2455$.

Figure 20.7 shows the behavior of the objective function with respect to the
number of iterations. In this case, the network was initiated with random values
uniformly distributed between zero and one.

These results show the effectiveness of the application of Hopfield networks to
solve constrained nonlinear optimization problems. The network was also simulated
considering different initial values assigned to the output vector v. For all simulations, the network always converged to the same solutions.

Although the number of iterations is relatively high, the time spent in obtaining
the solutions is quite short. Regarding this aspect, it must be mentioned that, though
sequential computers were used in this simulations, the network performance could

Fig. 20.7 Evolution of the
objective function (third
problem)

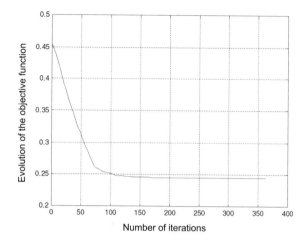

achieve maximum efficiency when it runs on computers with parallel processors. In such case, each processing unit corresponds to a neural unit.

Other important aspect concerns the quality of solutions obtained by the network. Despite that the solution values show some deviations from the optimal values, it can be observed that these deviations are smaller than the deviations obtained by the use of other neural approaches for solving these kinds of problems.

It is also noteworthy that some tests performed in bigger dimensional problems, having about 800 inequality constraints, showed that the network can become slow in its search for equilibrium points that represents the solution. This behavior is because all simulations were done on sequential computers. The use of machines with paralleled processing, as already mentioned, can overcome this situation.

Finally, the mapping of constrained nonlinear optimization problems by Hopfield networks has some particularities in comparison to primal methods (Bazaraa et al. 2006) used for solving these problems. These particularities are highlighted as follows: (i) no need to calculate the active set of constraints at each iteration; (ii) the network initialization vector (initial solution) does not need to belong to the feasible set defined by the constraints; (iii) no need to obtain, in each iteration, an acceptable search direction; and (iv) search mechanism does not depend on the use of Lagrange multipliers.

Appendix A

Training set relating to Sect. 3.6

Sample	x_1	x_2	x_3	d
1	−0.6508	0.1097	4.0009	−1.0000
2	−1.4492	0.8896	4.4005	−1.0000
3	2.0850	0.6876	12.0710	−1.0000
4	0.2626	1.1476	7.7985	1.0000
5	0.6418	1.0234	7.0427	1.0000
6	0.2569	0.6730	8.3265	−1.0000
7	1.1155	0.6043	7.4446	1.0000
8	0.0914	0.3399	7.0677	−1.0000
9	0.0121	0.5256	4.6316	1.0000
10	−0.0429	0.4660	5.4323	1.0000
11	0.4340	0.6870	8.2287	−1.0000
12	0.2735	1.0287	7.1934	1.0000
13	0.4839	0.4851	7.4850	−1.0000
14	0.4089	−0.1267	5.5019	−1.0000
15	1.4391	0.1614	8.5843	−1.0000
16	−0.9115	−0.1973	2.1962	−1.0000
17	0.3654	1.0475	7.4858	1.0000
18	0.2144	0.7515	7.1699	1.0000
19	0.2013	1.0014	6.5489	1.0000
20	0.6483	0.2183	5.8991	1.0000
21	−0.1147	0.2242	7.2435	−1.0000
22	−0.7970	0.8795	3.8762	1.0000
23	−1.0625	0.6366	2.4707	1.0000
24	0.5307	0.1285	5.6883	1.0000
25	−1.2200	0.7777	1.7252	1.0000
26	0.3957	0.1076	5.6623	−1.0000
27	−0.1013	0.5989	7.1812	−1.0000
28	2.4482	0.9455	11.2095	1.0000

(continued)

© Springer International Publishing Switzerland 2017
I.N. da Silva et al., *Artificial Neural Networks*,
DOI 10.1007/978-3-319-43162-8

(continued)

Sample	x_1	x_2	x_3	d
29	2.0149	0.6192	10.9263	−1.0000
30	0.2012	0.2611	5.4631	1.0000

Appendix B

Training set relating to Sect. 4.6

Sample	x_1	x_2	x_3	x_4	d
1	0.4329	−1.3719	0.7022	−0.8535	1.0000
2	0.3024	0.2286	0.8630	2.7909	−1.0000
3	0.1349	−0.6445	1.0530	0.5687	−1.0000
4	0.3374	−1.7163	0.3670	−0.6283	−1.0000
5	1.1434	−0.0485	0.6637	1.2606	1.0000
6	1.3749	−0.5071	0.4464	1.3009	1.0000
7	0.7221	−0.7587	0.7681	−0.5592	1.0000
8	0.4403	−0.8072	0.5154	−0.3129	1.0000
9	−0.5231	0.3548	0.2538	1.5776	−1.0000
10	0.3255	−2.0000	0.7112	−1.1209	1.0000
11	0.5824	1.3915	−0.2291	4.1735	−1.0000
12	0.1340	0.6081	0.4450	3.2230	−1.0000
13	0.1480	−0.2988	0.4778	0.8649	1.0000
14	0.7359	0.1869	−0.0872	2.3584	1.0000
15	0.7115	−1.1469	0.3394	0.9573	−1.0000
16	0.8251	−1.2840	0.8452	1.2382	−1.0000
17	0.1569	0.3712	0.8825	1.7633	1.0000
18	0.0033	0.6835	0.5389	2.8249	−1.0000
19	0.4243	0.8313	0.2634	3.5855	−1.0000
20	1.0490	0.1326	0.9138	1.9792	1.0000
21	1.4276	0.5331	−0.0145	3.7286	1.0000
22	0.5971	1.4865	0.2904	4.6069	−1.0000
23	0.8475	2.1479	0.3179	5.8235	−1.0000
24	1.3967	−0.4171	0.6443	1.3927	1.0000
25	0.0044	1.5378	0.6099	4.7755	−1.0000
26	0.2201	−0.5668	0.0515	0.7829	1.0000
27	0.6300	−1.2480	0.8591	0.8093	−1.0000
28	−0.2479	0.8960	0.0547	1.7381	1.0000

(continued)

© Springer International Publishing Switzerland 2017
I.N. da Silva et al., *Artificial Neural Networks*,
DOI 10.1007/978-3-319-43162-8

(continued)

29	−0.3088	−0.0929	0.8659	1.5483	−1.0000
30	−0.5180	1.4974	0.5453	2.3993	1.0000
31	0.6833	0.8266	0.0829	2.8864	1.0000
32	0.4353	−1.4066	0.4207	−0.4879	1.0000
33	−0.1069	−3.2329	0.1856	−2.4572	−1.0000
34	0.4662	0.6261	0.7304	3.4370	−1.0000
35	0.8298	−1.4089	0.3119	1.3235	−1.0000

Appendix C

Training set relating to Sect. 5.8

Sample	x_1	x_2	x_3	d	Sample	x_1	x_2	x_3	d
1	0.8799	0.7998	0.3972	0.8399	101	0.8203	0.0682	0.4260	0.5643
2	0.5700	0.5111	0.2418	0.6258	102	0.6226	0.2146	0.1021	0.4452
3	0.6796	0.4117	0.3370	0.6622	103	0.4589	0.3147	0.2236	0.4962
4	0.3567	0.2967	0.6037	0.5969	104	0.3471	0.8889	0.1564	0.5875
5	0.3866	0.8390	0.0232	0.5316	105	0.5762	0.8292	0.4116	0.7853
6	0.0271	0.7788	0.7445	0.6335	106	0.9053	0.6245	0.5264	0.8506
7	0.8174	0.8422	0.3229	0.8068	107	0.2860	0.0793	0.0549	0.2224
8	0.6027	0.1468	0.3759	0.5342	108	0.9567	0.3034	0.4425	0.6993
9	0.1203	0.3260	0.5419	0.4768	109	0.5170	0.9266	0.1565	0.6594
10	0.1325	0.2082	0.4934	0.4105	110	0.8149	0.0396	0.6227	0.6165
11	0.6950	1.0000	0.4321	0.8404	111	0.3710	0.3554	0.5633	0.6171
12	0.0036	0.1940	0.3274	0.2697	112	0.8702	0.3185	0.2762	0.6287
13	0.2650	0.0161	0.5947	0.4125	113	0.1016	0.6382	0.3173	0.4957
14	0.5849	0.6019	0.4376	0.7464	114	0.3890	0.2369	0.0083	0.3235
15	0.0108	0.3538	0.1810	0.2800	115	0.2702	0.8617	0.1218	0.5319
16	0.9008	0.7264	0.9184	0.9602	116	0.7473	0.6507	0.5582	0.8464
17	0.0023	0.9659	0.3182	0.4986	117	0.9108	0.2139	0.4641	0.6625
18	0.1366	0.6357	0.6967	0.6459	118	0.4343	0.6028	0.1344	0.5546
19	0.8621	0.7353	0.2742	0.7718	119	0.6847	0.4062	0.9318	0.8204
20	0.0682	0.9624	0.4211	0.5764	120	0.8657	0.9448	0.9900	0.9904
21	0.6112	0.6014	0.5254	0.7868	121	0.4011	0.4138	0.8715	0.7222
22	0.0030	0.7585	0.8928	0.6388	122	0.5949	0.2600	0.0810	0.4480
23	0.7644	0.5964	0.0407	0.6055	123	0.1845	0.7906	0.9725	0.7425
24	0.6441	0.2097	0.5847	0.6545	124	0.3438	0.6725	0.9821	0.7926
25	0.0803	0.3799	0.6020	0.4991	125	0.8398	0.1360	0.9119	0.7222
26	0.1908	0.8046	0.5402	0.6665	126	0.2245	0.0971	0.6136	0.4402
27	0.6937	0.3967	0.6055	0.7595	127	0.3742	0.9668	0.8194	0.8371
28	0.2591	0.0582	0.3978	0.3604	128	0.9572	0.9836	0.3793	0.8556

(continued)

© Springer International Publishing Switzerland 2017
I.N. da Silva et al., *Artificial Neural Networks*,
DOI 10.1007/978-3-319-43162-8

(continued)

Sample	x_1	x_2	x_3	d	Sample	x_1	x_2	x_3	d
29	0.4241	0.1850	0.9066	0.6298	129	0.7496	0.0410	0.1360	0.4059
30	0.3332	0.9303	0.2475	0.6287	130	0.9123	0.3510	0.0682	0.5455
31	0.3625	0.1592	0.9981	0.5948	131	0.6954	0.5500	0.6801	0.8388
32	0.9259	0.0960	0.1645	0.4716	132	0.5252	0.6529	0.5729	0.7893
33	0.8606	0.6779	0.0033	0.6242	133	0.3156	0.3851	0.5983	0.6161
34	0.0838	0.5472	0.3758	0.4835	134	0.1460	0.1637	0.0249	0.1813
35	0.0303	0.9191	0.7233	0.6491	135	0.7780	0.4491	0.4614	0.7498
36	0.9293	0.8319	0.9664	0.9840	136	0.5959	0.8647	0.8601	0.9176
37	0.7268	0.1440	0.9753	0.7096	137	0.2204	0.1785	0.4607	0.4276
38	0.2888	0.6593	0.4078	0.6328	138	0.7355	0.8264	0.7015	0.9214
39	0.5515	0.1364	0.2894	0.4745	139	0.9931	0.6727	0.3139	0.7829
40	0.7683	0.0067	0.5546	0.5708	140	0.9123	0.0000	0.1106	0.3944
41	0.6462	0.6761	0.8340	0.8933	141	0.2858	0.9688	0.2262	0.5988
42	0.3694	0.2212	0.1233	0.3658	142	0.7931	0.8993	0.9028	0.9728
43	0.2706	0.3222	0.9996	0.6310	143	0.7841	0.0778	0.9012	0.6832
44	0.6282	0.1404	0.8474	0.6733	144	0.1380	0.5881	0.2367	0.4622
45	0.5861	0.6693	0.3818	0.7433	145	0.6345	0.5165	0.7139	0.8191
46	0.6057	0.9901	0.5141	0.8466	146	0.2453	0.5888	0.1559	0.4765
47	0.5915	0.5588	0.3055	0.6787	147	0.1174	0.5436	0.3657	0.4953
48	0.8359	0.4145	0.5016	0.7597	148	0.3667	0.3228	0.6952	0.6376
49	0.5497	0.6319	0.8382	0.8521	149	0.9532	0.6949	0.4451	0.8426
50	0.7072	0.1721	0.3812	0.5772	150	0.7954	0.8346	0.0449	0.6676
51	0.1185	0.5084	0.8376	0.6211	151	0.1427	0.0480	0.6267	0.3780
52	0.6365	0.5562	0.4965	0.7693	152	0.1516	0.9824	0.0827	0.4627
53	0.4145	0.5797	0.8599	0.7878	153	0.4868	0.6223	0.7462	0.8116
54	0.2575	0.5358	0.4028	0.5777	154	0.3408	0.5115	0.0783	0.4559
55	0.2026	0.3300	0.3054	0.4261	155	0.8146	0.6378	0.5837	0.8628
56	0.3385	0.0476	0.5941	0.4625	156	0.2820	0.5409	0.7256	0.6939
57	0.4094	0.1726	0.7803	0.6015	157	0.5716	0.2958	0.5477	0.6619
58	0.1261	0.6181	0.4927	0.5739	158	0.9323	0.0229	0.4797	0.5731
59	0.1224	0.4662	0.2146	0.4007	159	0.2907	0.7245	0.5165	0.6911
60	0.6793	0.6774	1.0000	0.9141	160	0.0068	0.0545	0.0861	0.0851
61	0.8176	0.0358	0.2506	0.4707	161	0.2636	0.9885	0.2175	0.5847
62	0.6937	0.6685	0.5075	0.8220	162	0.0350	0.3653	0.7801	0.5117
63	0.2404	0.5411	0.8754	0.6980	163	0.9670	0.3031	0.7127	0.7836
64	0.6553	0.2609	0.1188	0.4851	164	0.0000	0.7763	0.8735	0.6388
65	0.8886	0.0288	0.2604	0.4802	165	0.4395	0.0501	0.9761	0.5712
66	0.3974	0.5275	0.6457	0.7215	166	0.9359	0.0366	0.9514	0.6826
67	0.2108	0.4910	0.5432	0.5913	167	0.0173	0.9548	0.4289	0.5527
68	0.8675	0.5571	0.1849	0.6805	168	0.6112	0.9070	0.6286	0.8803

(continued)

(continued)

Sample	x_1	x_2	x_3	d	Sample	x_1	x_2	x_3	d
69	0.5693	0.0242	0.9293	0.6033	169	0.2010	0.9573	0.6791	0.7283
70	0.8439	0.4631	0.6345	0.8226	170	0.8914	0.9144	0.2641	0.7966
71	0.3644	0.2948	0.3937	0.5240	171	0.0061	0.0802	0.8621	0.3711
72	0.2014	0.6326	0.9782	0.7143	172	0.2212	0.4664	0.3821	0.5260
73	0.4039	0.0645	0.4629	0.4547	173	0.2401	0.6964	0.0751	0.4637
74	0.7137	0.0670	0.2359	0.4602	174	0.7881	0.9833	0.3038	0.8049
75	0.4277	0.9555	0.0000	0.5477	175	0.2435	0.0794	0.5551	0.4223
76	0.0259	0.7634	0.2889	0.4738	176	0.2752	0.8414	0.2797	0.6079
77	0.1871	0.7682	0.9697	0.7397	177	0.7616	0.4698	0.5337	0.7809
78	0.3216	0.5420	0.0677	0.4526	178	0.3395	0.0022	0.0087	0.1836
79	0.2524	0.7688	0.9523	0.7711	179	0.7849	0.9981	0.4449	0.8641
80	0.3621	0.5295	0.2521	0.5571	180	0.8312	0.0961	0.2129	0.4857
81	0.2942	0.1625	0.2745	0.3759	181	0.9763	0.1102	0.6227	0.6667
82	0.8180	0.0023	0.1439	0.4018	182	0.8597	0.3284	0.6932	0.7829
83	0.8429	0.1704	0.5251	0.6563	183	0.9295	0.3275	0.7536	0.8016
84	0.9612	0.6898	0.6630	0.9128	184	0.2435	0.2163	0.7625	0.5449
85	0.1009	0.4190	0.0826	0.3055	185	0.9281	0.8356	0.5285	0.8991
86	0.7071	0.7704	0.8328	0.9298	186	0.8313	0.7566	0.6192	0.9047
87	0.3371	0.7819	0.0959	0.5377	187	0.1712	0.0545	0.5033	0.3561
88	0.1555	0.5599	0.9221	0.6663	188	0.0609	0.1702	0.4306	0.3310
89	0.7318	0.1877	0.3311	0.5689	189	0.5899	0.9408	0.0369	0.6245
90	0.1665	0.7449	0.0997	0.4508	190	0.7858	0.5115	0.0916	0.6066
91	0.8762	0.2498	0.9167	0.7829	191	1.0000	0.1653	0.7103	0.7172
92	0.9885	0.6229	0.2085	0.7200	192	0.2007	0.1163	0.3431	0.3385
93	0.0461	0.7745	0.5632	0.5949	193	0.2306	0.0330	0.0293	0.1590
94	0.3209	0.6229	0.5233	0.6810	194	0.8477	0.6378	0.4623	0.8254
95	0.9189	0.5930	0.7288	0.8989	195	0.9677	0.7895	0.9467	0.9782
96	0.0382	0.5515	0.8818	0.5999	196	0.0339	0.4669	0.1526	0.3250
97	0.3726	0.9988	0.3814	0.7086	197	0.0080	0.8988	0.4201	0.5404
98	0.4211	0.2668	0.3307	0.5080	198	0.9955	0.8897	0.6175	0.9360
99	0.2378	0.0817	0.3574	0.3452	199	0.7408	0.5351	0.2732	0.6949
100	0.9893	0.7637	0.2526	0.7755	200	0.6843	0.3737	0.1562	0.5625

Training set relating to Sect. 5.9

Sample	x_1	x_2	x_3	x_4	d_1	d_2	d_3
1	0.3841	0.2021	0.0000	0.2438	1	0	0
2	0.1765	0.1613	0.3401	0.0843	1	0	0
3	0.3170	0.5786	0.3387	0.4192	0	1	0
4	0.2467	0.0337	0.2699	0.3454	1	0	0
5	0.6102	0.8192	0.4679	0.4762	0	1	0
6	0.7030	0.7784	0.7482	0.6562	0	0	1
7	0.4767	0.4348	0.4852	0.3640	0	1	0
8	0.7589	0.8256	0.6514	0.6143	0	0	1
9	0.1579	0.3641	0.2551	0.2919	1	0	0
10	0.5561	0.5602	0.5605	0.2105	0	1	0
11	0.3267	0.2974	0.0343	0.1466	1	0	0
12	0.2303	0.0942	0.3889	0.1713	1	0	0
13	0.2953	0.2963	0.2600	0.3039	1	0	0
14	0.5797	0.4789	0.5780	0.3048	0	1	0
15	0.5860	0.5250	0.4792	0.4021	0	1	0
16	0.7045	0.6933	0.6449	0.6623	0	0	1
17	0.9134	0.9412	0.6078	0.5934	0	0	1
18	0.2333	0.4943	0.2525	0.2567	1	0	0
19	0.2676	0.4172	0.2775	0.2721	1	0	0
20	0.4850	0.5506	0.5269	0.6036	0	1	0
21	0.2434	0.2567	0.2312	0.2624	1	0	0
22	0.1250	0.3023	0.1826	0.3168	1	0	0
23	0.5598	0.4253	0.4258	0.3192	0	1	0
24	0.5738	0.7674	0.6154	0.4447	0	0	1

Sample	x_1	x_2	x_3	x_4	d_1	d_2	d_3
66	0.3367	0.4333	0.2336	0.1678	1	0	0
67	0.4744	0.4604	0.1507	0.4873	1	0	0
68	0.7510	0.4350	0.5453	0.4831	0	1	0
69	0.4045	0.5636	0.2534	0.5573	0	1	0
70	0.1449	0.1539	0.2446	0.0559	1	0	0
71	0.3460	0.2722	0.1866	0.5049	1	0	0
72	0.2241	0.2046	0.3575	0.2891	1	0	0
73	0.1412	0.2264	0.4025	0.2661	1	0	0
74	0.5782	0.6418	0.7212	0.6396	0	0	1
75	0.9153	0.6571	0.8229	0.6689	0	0	1
76	0.6014	0.7664	0.6385	0.5513	0	0	1
77	0.7328	0.8708	0.8812	0.7060	0	0	1
78	0.4270	0.6352	0.6811	0.3884	0	1	0
79	0.6189	0.1652	0.4016	0.3042	1	0	0
80	0.2143	0.3868	0.1926	0.0000	1	0	0
81	0.5696	0.7238	0.7199	0.6677	0	0	1
82	0.8656	0.6700	0.6570	0.6065	0	0	1
83	0.9002	0.6858	0.7409	0.7047	0	0	1
84	0.4167	0.5255	0.5506	0.4093	0	1	0
85	0.8325	0.4804	0.7990	0.7471	0	0	1
86	0.4124	0.1191	0.4720	0.3184	1	0	0
87	1.0000	1.0000	0.7924	0.7074	0	0	1
88	0.5685	0.6924	0.6180	0.5792	0	1	0
89	0.6505	0.4864	0.2972	0.4599	0	1	0

(continued)

(continued)

Sample	x_1	x_2	x_3	x_4	d_1	d_2	d_3
90	0.8124	0.7690	0.9720	1.0000	0	0	1
91	0.9013	0.7160	1.0000	0.8046	0	0	1
92	0.8872	0.7556	0.9307	0.6791	0	0	1
93	0.3708	0.2139	0.2136	0.4295	1	0	0
94	0.5159	0.4349	0.3715	0.4086	0	0	0
95	0.6768	0.6304	0.8044	0.4885	0	1	0
96	0.1664	0.2404	0.2000	0.3425	0	0	1
97	0.2495	0.2807	0.4679	0.2200	1	0	0
98	0.2487	0.2348	0.0913	0.1281	1	0	0
99	0.5748	0.8552	0.5973	0.7317	0	0	1
100	0.3858	0.7585	0.3239	0.3565	0	1	0
101	0.3329	0.4946	0.5614	0.3152	0	1	0
102	0.3891	0.4805	0.7598	0.4231	0	1	0
103	0.2888	0.4888	0.1930	0.0177	1	1	0
104	0.3827	0.4900	0.2272	0.3599	0	1	0
105	0.6047	0.4224	0.6274	0.5809	0	1	0
106	0.9840	0.7031	0.6469	0.4701	0	0	0
107	0.6554	0.6785	0.9279	0.7723	0	0	1
108	0.0466	0.3388	0.0840	0.0762	0	0	1
109	0.6154	0.8196	0.6339	0.7729	1	0	0
110	0.8452	0.8897	0.8383	0.6961	0	0	1
111	0.6927	0.7870	0.7689	0.7213	0	0	1
112	0.4032	0.6188	0.4930	0.5380	0	1	1
113	0.4006	0.3094	0.3868	0.0811	1	0	0

(continued)

Sample	x_1	x_2	x_3	x_4	d_1	d_2	d_3
25	0.5692	0.8368	0.5832	0.4585	0	0	1
26	0.4655	0.7682	0.3221	0.2940	0	1	0
27	0.5568	0.7592	0.6293	0.5453	0	1	0
28	0.8842	0.7509	0.5723	0.5814	0	0	1
29	0.7959	0.9243	0.7339	0.7334	0	0	1
30	0.7124	0.7128	0.6065	0.6668	0	0	1
31	0.6749	0.8767	0.6543	0.7461	0	0	1
32	0.3674	0.4359	0.4230	0.2965	1	0	0
33	0.3473	0.0754	0.2183	0.1905	1	0	0
34	0.6931	0.5188	0.5386	0.5794	0	1	0
35	0.6439	0.4959	0.4322	0.4582	0	1	0
36	0.5627	0.4893	0.6831	0.5120	0	1	0
37	0.5182	0.7553	0.6368	0.4538	0	1	0
38	0.6046	0.7479	0.6542	0.4375	0	1	0
39	0.6328	0.6786	0.7751	0.6183	0	0	1
40	0.3429	0.4694	0.2855	0.2977	1	0	0
41	0.6371	0.5069	0.5316	0.4520	0	1	0
42	0.6388	0.6970	0.6407	0.7677	0	0	1
43	0.3529	0.5504	0.3706	0.4828	0	1	0
44	0.4302	0.3237	0.6397	0.4319	1	1	0
45	0.7078	0.9604	0.7470	0.6399	0	0	1
46	0.7350	0.8170	0.7227	0.6279	0	0	1
47	0.7011	0.2946	0.6625	0.4312	0	1	0
48	0.5961	0.3817	0.6363	0.3663	0	1	0

(continued)

(continued)

Sample	x_1	x_2	x_3	x_4	d_1	d_2	d_3
49	0.0000	0.2563	0.2603	0.3027	1	0	0
50	0.5996	0.5704	0.6965	0.6548	0	0	1
51	0.4289	0.3709	0.3994	0.3656	0	1	0
52	0.2093	0.3655	0.3334	0.1802	1	0	0
53	0.2335	0.2856	0.3912	0.1601	1	0	0
54	0.3266	0.7751	0.4356	0.3448	0	1	0
55	0.2457	0.1203	0.1228	0.2206	1	0	0
56	0.4656	0.4815	0.4211	0.4862	0	1	0
57	0.7511	0.8868	0.5408	0.6253	0	0	1
58	0.7825	0.9386	0.6510	0.6996	0	0	1
59	0.3463	0.4118	0.2507	0.0454	1	0	0
60	0.5172	0.1482	0.3172	0.2323	1	0	0
61	0.6942	0.4516	0.5387	0.5983	0	1	0
62	0.7586	0.7017	0.7120	0.7509	0	0	1
63	0.6880	0.6004	0.6602	0.4320	0	1	0
64	0.4742	0.5079	0.4135	0.4161	0	1	0
65	0.4419	0.5761	0.4515	0.4497	0	1	0

Sample	x_1	x_2	x_3	x_4	d_1	d_2	d_3
114	0.7416	0.7138	0.6823	0.6067	0	0	1
115	0.7404	0.6764	0.8293	0.4694	0	0	1
116	0.7736	0.7097	0.6826	0.8142	0	0	1
117	0.5823	0.9635	0.3706	0.5636	0	1	0
118	0.2081	0.3738	0.3119	0.3552	1	0	0
119	0.5616	0.8972	0.5186	0.6650	0	0	1
120	0.6594	0.8907	0.6000	0.7157	0	0	1
121	0.3979	0.3070	0.3637	0.1220	1	0	0
122	0.2644	0.0000	0.3572	0.1931	1	0	0
123	0.4816	0.4791	0.4213	0.5889	0	1	0
124	0.0848	0.0749	0.4349	0.3328	1	0	0
125	0.4608	0.6775	0.3533	0.3016	0	1	0
126	0.4155	0.6589	0.5310	0.5404	0	1	0
127	0.3934	0.6244	0.4817	0.4324	0	1	0
128	0.5843	0.8517	0.8576	0.7133	0	0	1
129	0.1995	0.3690	0.3537	0.3462	1	0	0
130	0.3832	0.2321	0.0341	0.2450	1	0	0

Training set relating to Sect. 5.10

Sample	$f(t)$	Sample	$f(t)$	Sample	$f(t)$	Sample	$f(t)$
$t = 1$	0.1701	$t = 26$	0.2398	$t = 51$	0.3087	$t = 76$	0.3701
$t = 2$	0.1023	$t = 27$	0.0508	$t = 52$	0.0159	$t = 77$	0.0006
$t = 3$	0.4405	$t = 28$	0.4497	$t = 53$	0.4330	$t = 78$	0.3943
$t = 4$	0.3609	$t = 29$	0.2178	$t = 54$	0.0733	$t = 79$	0.0646
$t = 5$	0.7192	$t = 30$	0.7762	$t = 55$	0.7995	$t = 80$	0.7878
$t = 6$	0.2258	$t = 31$	0.1078	$t = 56$	0.0262	$t = 81$	0.1694
$t = 7$	0.3175	$t = 32$	0.3773	$t = 57$	0.4223	$t = 82$	0.4468
$t = 8$	0.0127	$t = 33$	0.0001	$t = 58$	0.0085	$t = 83$	0.0372
$t = 9$	0.4290	$t = 34$	0.3877	$t = 59$	0.3303	$t = 84$	0.2632
$t = 10$	0.0544	$t = 35$	0.0821	$t = 60$	0.2037	$t = 85$	0.3048
$t = 11$	0.8000	$t = 36$	0.7836	$t = 61$	0.7332	$t = 86$	0.6516
$t = 12$	0.0450	$t = 37$	0.1887	$t = 62$	0.3328	$t = 87$	0.4690
$t = 13$	0.4268	$t = 38$	0.4483	$t = 63$	0.4445	$t = 88$	0.4132
$t = 14$	0.0112	$t = 39$	0.0424	$t = 64$	0.0909	$t = 89$	0.1523
$t = 15$	0.3218	$t = 40$	0.2539	$t = 65$	0.1838	$t = 90$	0.1182
$t = 16$	0.2185	$t = 41$	0.3164	$t = 66$	0.3888	$t = 91$	0.4334
$t = 17$	0.7240	$t = 42$	0.6386	$t = 67$	0.5277	$t = 92$	0.3978
$t = 18$	0.3516	$t = 43$	0.4862	$t = 68$	0.6042	$t = 93$	0.6987
$t = 19$	0.4420	$t = 44$	0.4068	$t = 69$	0.3435	$t = 94$	0.2538
$t = 20$	0.0984	$t = 45$	0.1611	$t = 70$	0.2304	$t = 95$	0.2998
$t = 21$	0.1747	$t = 46$	0.1101	$t = 71$	0.0568	$t = 96$	0.0195
$t = 22$	0.3964	$t = 47$	0.4372	$t = 72$	0.4500	$t = 97$	0.4366
$t = 23$	0.5114	$t = 48$	0.3795	$t = 73$	0.2371	$t = 98$	0.0924
$t = 24$	0.6183	$t = 49$	0.7092	$t = 74$	0.7705	$t = 99$	0.7984
$t = 25$	0.3330	$t = 50$	0.2400	$t = 75$	0.1246	$t = 100$	0.0077

Appendix D

Training set relating to Sect. 6.5

Sample	x_1	x_2	x_3	d	Sample	x_1	x_2	x_3	d
1	0.9532	0.6949	0.4451	0.8426	76	0.6441	0.2097	0.5847	0.6545
2	0.7954	0.8346	0.0449	0.6676	77	0.0803	0.3799	0.6020	0.4991
3	0.1427	0.048	0.6267	0.3780	78	0.1908	0.8046	0.5402	0.6665
4	0.1516	0.9824	0.0827	0.4627	79	0.6937	0.3967	0.6055	0.7595
5	0.4868	0.6223	0.7462	0.8116	80	0.2591	0.0582	0.3978	0.3604
6	0.3408	0.5115	0.0783	0.4559	81	0.4241	0.1850	0.9066	0.6298
7	0.8146	0.6378	0.5837	0.8628	82	0.3332	0.9303	0.2475	0.6287
8	0.2820	0.5409	0.7256	0.6939	83	0.3625	0.1592	0.9981	0.5948
9	0.5716	0.2958	0.5477	0.6619	84	0.9259	0.0960	0.1645	0.4716
10	0.9323	0.0229	0.4797	0.5731	85	0.8606	0.6779	0.0033	0.6242
11	0.2907	0.7245	0.5165	0.6911	86	0.0838	0.5472	0.3758	0.4835
12	0.0068	0.0545	0.0861	0.0851	87	0.0303	0.9191	0.7233	0.6491
13	0.2636	0.9885	0.2175	0.5847	88	0.9293	0.8319	0.9664	0.9840
14	0.035	0.3653	0.7801	0.5117	89	0.7268	0.1440	0.9753	0.7096
15	0.967	0.3031	0.7127	0.7836	90	0.2888	0.6593	0.4078	0.6328
16	0.0000	0.7763	0.8735	0.6388	91	0.5515	0.1364	0.2894	0.4745
17	0.4395	0.0501	0.9761	0.5712	92	0.7683	0.0067	0.5546	0.5708
18	0.9359	0.0366	0.9514	0.6826	93	0.6462	0.6761	0.8340	0.8933
19	0.0173	0.9548	0.4289	0.5527	94	0.3694	0.2212	0.1233	0.3658
20	0.6112	0.907	0.6286	0.8803	95	0.2706	0.3222	0.9996	0.6310
21	0.2010	0.9573	0.6791	0.7283	96	0.6282	0.1404	0.8474	0.6733
22	0.8914	0.9144	0.2641	0.7966	97	0.5861	0.6693	0.3818	0.7433
23	0.0061	0.0802	0.8621	0.3711	98	0.6057	0.9901	0.5141	0.8466
24	0.2212	0.4664	0.3821	0.5260	99	0.5915	0.5588	0.3055	0.6787
25	0.2401	0.6964	0.0751	0.4637	100	0.8359	0.4145	0.5016	0.7597
26	0.7881	0.9833	0.3038	0.8049	101	0.5497	0.6319	0.8382	0.8521
27	0.2435	0.0794	0.5551	0.4223	102	0.7072	0.1721	0.3812	0.5772
28	0.2752	0.8414	0.2797	0.6079	103	0.1185	0.5084	0.8376	0.6211

(continued)

© Springer International Publishing Switzerland 2017
I.N. da Silva et al., *Artificial Neural Networks*,
DOI 10.1007/978-3-319-43162-8

(continued)

Sample	x_1	x_2	x_3	d	Sample	x_1	x_2	x_3	d
29	0.7616	0.4698	0.5337	0.7809	104	0.6365	0.5562	0.4965	0.7693
30	0.3395	0.0022	0.0087	0.1836	105	0.4145	0.5797	0.8599	0.7878
31	0.7849	0.9981	0.4449	0.8641	106	0.2575	0.5358	0.4028	0.5777
32	0.8312	0.0961	0.2129	0.4857	107	0.2026	0.3300	0.3054	0.4261
33	0.9763	0.1102	0.6227	0.6667	108	0.3385	0.0476	0.5941	0.4625
34	0.8597	0.3284	0.6932	0.7829	109	0.4094	0.1726	0.7803	0.6015
35	0.9295	0.3275	0.7536	0.8016	110	0.1261	0.6181	0.4927	0.5739
36	0.2435	0.2163	0.7625	0.5449	111	0.1224	0.4662	0.2146	0.4007
37	0.9281	0.8356	0.5285	0.8991	112	0.6793	0.6774	1.0000	0.9141
38	0.8313	0.7566	0.6192	0.9047	113	0.8176	0.0358	0.2506	0.4707
39	0.1712	0.0545	0.5033	0.3561	114	0.6937	0.6685	0.5075	0.8220
40	0.0609	0.1702	0.4306	0.3310	115	0.2404	0.5411	0.8754	0.6980
41	0.5899	0.9408	0.0369	0.6245	116	0.6553	0.2609	0.1188	0.4851
42	0.7858	0.5115	0.0916	0.6066	117	0.8886	0.0288	0.2604	0.4802
43	1.0000	0.1653	0.7103	0.7172	118	0.3974	0.5275	0.6457	0.7215
44	0.2007	0.1163	0.3431	0.3385	119	0.2108	0.4910	0.5432	0.5913
45	0.2306	0.033	0.0293	0.1590	120	0.8675	0.5571	0.1849	0.6805
46	0.8477	0.6378	0.4623	0.8254	121	0.5693	0.0242	0.9293	0.6033
47	0.9677	0.7895	0.9467	0.9782	122	0.8439	0.4631	0.6345	0.8226
48	0.0339	0.4669	0.1526	0.3250	123	0.3644	0.2948	0.3937	0.5240
49	0.008	0.8988	0.4201	0.5404	124	0.2014	0.6326	0.9782	0.7143
50	0.9955	0.8897	0.6175	0.9360	125	0.4039	0.0645	0.4629	0.4547
51	0.7408	0.5351	0.2732	0.6949	126	0.7137	0.0670	0.2359	0.4602
52	0.6843	0.3737	0.1562	0.5625	127	0.4277	0.9555	0.0000	0.5477
53	0.8799	0.7998	0.3972	0.8399	128	0.0259	0.7634	0.2889	0.4738
54	0.5700	0.5111	0.2418	0.6258	129	0.1871	0.7682	0.9697	0.7397
55	0.6796	0.4117	0.3370	0.6622	130	0.3216	0.5420	0.0677	0.4526
56	0.3567	0.2967	0.6037	0.5969	131	0.2524	0.7688	0.9523	0.7711
57	0.3866	0.8390	0.0232	0.5316	132	0.3621	0.5295	0.2521	0.5571
58	0.0271	0.7788	0.7445	0.6335	133	0.2942	0.1625	0.2745	0.3759
59	0.8174	0.8422	0.3229	0.8068	134	0.8180	0.0023	0.1439	0.4018
60	0.6027	0.1468	0.3759	0.5342	135	0.8429	0.1704	0.5251	0.6563
61	0.1203	0.3260	0.5419	0.4768	136	0.9612	0.6898	0.6630	0.9128
62	0.1325	0.2082	0.4934	0.4105	137	0.1009	0.419	0.0826	0.3055
63	0.6950	1.0000	0.4321	0.8404	138	0.7071	0.7704	0.8328	0.9298
64	0.0036	0.1940	0.3274	0.2697	139	0.3371	0.7819	0.0959	0.5377
65	0.2650	0.0161	0.5947	0.4125	140	0.9931	0.6727	0.3139	0.7829
66	0.5849	0.6019	0.4376	0.7464	141	0.9123	0.0000	0.1106	0.3944
67	0.0108	0.3538	0.1810	0.2800	142	0.2858	0.9688	0.2262	0.5988
68	0.9008	0.7264	0.9184	0.9602	143	0.7931	0.8993	0.9028	0.9728

(continued)

(continued)

Sample	x_1	x_2	x_3	d	Sample	x_1	x_2	x_3	d
69	0.0023	0.9659	0.3182	0.4986	144	0.7841	0.0778	0.9012	0.6832
70	0.1366	0.6357	0.6967	0.6459	145	0.1380	0.5881	0.2367	0.4622
71	0.8621	0.7353	0.2742	0.7718	146	0.6345	0.5165	0.7139	0.8191
72	0.0682	0.9624	0.4211	0.5764	147	0.2453	0.5888	0.1559	0.4765
73	0.6112	0.6014	0.5254	0.7868	148	0.1174	0.5436	0.3657	0.4953
74	0.0030	0.7585	0.8928	0.6388	149	0.3667	0.3228	0.6952	0.6376
75	0.7644	0.5964	0.0407	0.6055	150	0.2204	0.1785	0.4607	0.4276

Training set relating to Sect. 6.6

Sample	x_1	x_2	d	Sample	x_1	x_2	d
1	0.2563	0.9503	−1	21	0.456	0.1871	1
2	0.2405	0.9018	−1	22	0.1715	0.7713	1
3	0.1157	0.3676	1	23	0.5571	0.5485	−1
4	0.5147	0.0167	1	24	0.3344	0.0259	1
5	0.4127	0.3275	1	25	0.4803	0.7635	−1
6	0.2809	0.583	1	26	0.9721	0.485	−1
7	0.8263	0.9301	−1	27	0.8318	0.7844	−1
8	0.9359	0.8724	−1	28	0.1373	0.0292	1
9	0.1096	0.9165	−1	29	0.366	0.8581	−1
10	0.5158	0.8545	−1	30	0.3626	0.7302	−1
11	0.1334	0.1362	1	31	0.6474	0.3324	1
12	0.6371	0.1439	1	32	0.3461	0.2398	1
13	0.7052	0.6277	−1	33	0.1353	0.812	1
14	0.8703	0.8666	−1	34	0.3463	0.1017	1
15	0.2612	0.6109	1	35	0.9086	0.1947	−1
16	0.0244	0.5279	1	36	0.5227	0.2321	1
17	0.9588	0.3672	−1	37	0.5153	0.2041	1
18	0.9332	0.5499	−1	38	0.1832	0.0661	1
19	0.9623	0.2961	−1	39	0.5015	0.9812	−1
20	0.7297	0.5776	−1	40	0.5024	0.5274	−1

Appendix E

Training set relating to Sect. 8.5

Sample	x_1	x_2	x_3	Sample	x_1	x_2	x_3
1	0.2417	0.2857	0.2397	61	0.4856	0.6600	0.4798
2	0.2268	0.2874	0.2153	62	0.4114	0.7220	0.5106
3	0.1975	0.3315	0.1965	63	0.5671	0.7935	0.5929
4	0.3414	0.3166	0.1074	64	0.4875	0.7928	0.5532
5	0.2587	0.1918	0.2634	65	0.5172	0.7147	0.5774
6	0.2455	0.2075	0.1344	66	0.5483	0.6773	0.4842
7	0.3163	0.1679	0.1725	67	0.5740	0.6682	0.5335
8	0.2704	0.2605	0.1411	68	0.4587	0.6981	0.5900
9	0.1871	0.2965	0.1231	69	0.5794	0.7410	0.4759
10	0.3474	0.2715	0.1958	70	0.4712	0.6734	0.5677
11	0.2059	0.2928	0.2839	71	0.5126	0.8141	0.5224
12	0.2442	0.2272	0.2384	72	0.5557	0.7749	0.4342
13	0.2126	0.3437	0.1128	73	0.4916	0.8267	0.4586
14	0.2562	0.2542	0.1599	74	0.4629	0.8129	0.4950
15	0.1640	0.2289	0.2627	75	0.5850	0.7358	0.5107
16	0.2795	0.1880	0.1627	76	0.4435	0.7030	0.4594
17	0.3463	0.1513	0.2281	77	0.4155	0.7516	0.5524
18	0.3430	0.1508	0.1881	78	0.4887	0.7027	0.5886
19	0.1981	0.2821	0.1294	79	0.5462	0.7378	0.5107
20	0.2322	0.3025	0.2191	80	0.5251	0.8124	0.5686
21	0.7352	0.2722	0.6962	81	0.4635	0.7339	0.5638
22	0.7191	0.1825	0.7470	82	0.5907	0.7144	0.4718
23	0.6921	0.1537	0.8172	83	0.4982	0.8335	0.4597
24	0.6833	0.2048	0.8490	84	0.5242	0.7325	0.4079
25	0.8012	0.2684	0.7673	85	0.4075	0.8372	0.4271
26	0.7860	0.1734	0.7198	86	0.5934	0.8284	0.5107
27	0.7205	0.1542	0.7295	87	0.5463	0.6766	0.5639
28	0.6549	0.3288	0.8153	88	0.4403	0.8495	0.4806

(continued)

© Springer International Publishing Switzerland 2017
I.N. da Silva et al., *Artificial Neural Networks*,
DOI 10.1007/978-3-319-43162-8

(continued)

Sample	x_1	x_2	x_3	Sample	x_1	x_2	x_3
29	0.6968	0.3173	0.7389	89	0.4531	0.7760	0.5276
30	0.7448	0.2095	0.6847	90	0.5109	0.7387	0.5373
31	0.6746	0.3277	0.6725	91	0.5383	0.7780	0.4955
32	0.7897	0.2801	0.7679	92	0.5679	0.7156	0.5022
33	0.8399	0.3067	0.7003	93	0.5762	0.7781	0.5908
34	0.8065	0.3206	0.7205	94	0.5997	0.7504	0.5678
35	0.8357	0.3220	0.7879	95	0.4138	0.6975	0.5148
36	0.7438	0.3230	0.8384	96	0.5490	0.6674	0.4472
37	0.8172	0.3319	0.7628	97	0.4719	0.7527	0.4401
38	0.8248	0.2614	0.8405	98	0.4458	0.8063	0.4253
39	0.6979	0.2142	0.7309	99	0.4983	0.8131	0.5625
40	0.6804	0.3181	0.7017	100	0.5742	0.6789	0.5997
41	0.6973	0.3194	0.7522	101	0.5289	0.7354	0.4718
42	0.7910	0.2239	0.7018	102	0.5927	0.7738	0.5390
43	0.7052	0.2148	0.6866	103	0.5199	0.7131	0.4028
44	0.8088	0.1908	0.7563	104	0.5716	0.6558	0.4451
45	0.7640	0.1676	0.6994	105	0.5075	0.7045	0.4233
46	0.7616	0.2881	0.8087	106	0.4886	0.7004	0.4608
47	0.8188	0.2461	0.7273	107	0.5527	0.8243	0.5772
48	0.7920	0.3178	0.7497	108	0.4816	0.6969	0.4678
49	0.7802	0.1871	0.8102	109	0.5809	0.6557	0.4266
50	0.7332	0.2543	0.8194	110	0.5881	0.7565	0.4003
51	0.6921	0.1529	0.7759	111	0.5334	0.8446	0.4934
52	0.6833	0.2197	0.6943	112	0.4603	0.7992	0.4816
53	0.7860	0.1745	0.7639	113	0.5491	0.6504	0.4063
54	0.8009	0.3082	0.8491	114	0.4288	0.8455	0.5047
55	0.7793	0.1935	0.6738	115	0.5636	0.7884	0.5417
56	0.7373	0.2698	0.7864	116	0.5349	0.6736	0.4541
57	0.7048	0.2380	0.7825	117	0.5569	0.8393	0.5652
58	0.8393	0.2857	0.7733	118	0.4729	0.7702	0.5325
59	0.6878	0.2126	0.6961	119	0.5472	0.8454	0.5449
60	0.6651	0.3492	0.6737	120	0.5805	0.7349	0.4464

References

Agarwal, M., Agrawal, H., Jain, N., & Kumar, M. (2010). Face recognition using principle component analysis, eigenface and neural network. In *Proceedings of the IEEE International Conference on Signal Acquisition and Processing*, Bangalore, India, pp. 310–314.

Aihara, K., Takabe, T., & Toyoda, M. (1990). Chaotic neural networks. *Physics Letters A, 144*, 333–340.

Aiyer, S. V. B., Niranjan, M., & Fallside, F. (1990). A theoretical investigation into the perform of the Hopfield model. *IEEE Transactions on Neural Networks, 1*(2), 204–215.

Alligood, K. T., Sauer, T. D., & Yorke, J. A. (2009). *Chaos—An introduction to dynamical systems*. New York, USA: Springer.

Amit, D. J. (1992). *Modeling brain function—The world of attractor neural networks*. Cambridge, UK: Cambridge University Press.

Amit, D. J., Gutfreund, H., & Sompolinsky, H. (1985). Spin-glass models of neural networks. *Physical Review A, 32*, 1007–1018.

Argyria, A. A., Panagoua, E. Z., Tarantilisc, P. A., Polysiouc, M., & Nychasa, G.-J. E. (2010). Rapid qualitative and quantitative detection of beef fillets spoilage based on Fourier transform infrared spectroscopy data and artificial neural networks. *Sensors and Actuators B: Chemical, 145*(1), 146–154.

Armitage, D. W., & Ober, H. K. (2010). A comparison of supervised learning techniques in the classification of bat echolocation calls. *Ecological Informatics, 5*(6), 465–473.

Atencia, M., Joya, G., & Sandoval, F. (2005). Dynamical analysis of continuous higher-order Hopfield networks for combinatorial optimization. *Neural Computation, 17*, 1802–1819.

Aurenhammer, F. (1991). Voronoy diagrams—A survey of a fundamental geometric data structure. *ACM Computing Surveys, 23*(3), 345–405.

Baraldi, A., & Alpaydin, E. (2002). Constructive feedforward ART clustering networks. *IEEE Transactions on Neural Networks, 13*(3), 645–661.

Battiti, R. (1992). First and second order methods for learning: Between steepest descent and Newton's method. *Neural Computation, 4*(2), 141–166.

Bazaraa, M. S., Sherali, H. D., & Shetty, C. M. (2006). *Nonlinear programming—Theory and Algorithms* (3rd ed.). Hoboken, New Jersey, USA: Wiley.

Beiu, V., Quintana, J. M., & Avedillo, M. J. (2003). VLSI implementations of threshold logic—A comprehensive survey. *IEEE Transactions on Neural Networks, 14*(5), 1217–1243.

Bertsekas, D. P., & Tsitsiklis, J. N. (1996). *Neuro-dynamic programming*. Belmont, Massachusetts, USA: Athena Scientific.

Bowyer, A. (1981). Computing Dirichlet tessellations. *The Computer Journal, 24*(2), 162–166.

Boylestad, R., & Nashelsky, L. (2012). *Electronic devices and circuit theory* (11th ed.). Upper Saddle River, New Jersey, USA: Prentice-Hall.

Buckley, J. J., & Siler, W. (2004). *Fuzzy expert systems and fuzzy reasoning*. New York, USA: Wiley.

© Springer International Publishing Switzerland 2017
I.N. da Silva et al., *Artificial Neural Networks*,
DOI 10.1007/978-3-319-43162-8

Carpenter, G. A., & Grossberg, S. (1987). A massively parallel architecture for a self-organizing neural pattern recognition machine. *Computer Vision, Graphics, and Image Processing, 37*(1), 54–115.

Carpenter, G. A., & Grossberg, S. (1987). ART2: Self-organization of stable category recognition codes for analog input patterns. *Applied Optics, 26*(23), 4919–4930.

Carpenter, G. A., & Grossberg, S. (1988). The ART of adaptive pattern recognition by a self-organizing neural network. *Computer IEEE, 21*(3), 77–88.

Carpenter, G. A., & Grossberg, S. (1990). ART3: Hierarchical search using chemical transmitters in self-organizing pattern recognition architectures. *Neural Networks, 3*(2), 129–152.

Carpenter, G. A., Grossberg, S., & Reynolds, J. H. (1991). ARTMAP: Supervised real-time learning and classification of nonstationary data by a self-organizing neural network. *Neural Networks, 4*(5), 565–588.

Carpenter, G. A., Grossberg, S., & Rosen, D. B. (1991). Fuzzy ART: Fast stable learning and categorization of analog patterns by an adaptive resonance system. *Neural Networks, 4*(6), 759–771.

Castro, L. N., & Timmis, J. (2002). *Artificial immune systems—A new computational intelligence approach*. London, UK: Springer.

Chang, C. -Y, & Su, S.-J. (2005). The application of a full counter-propagation neural network to image watermarking. In *Proceedings of the IEEE International Conference on Networking, Sensing and Control* (pp. 993–998). Tucson, Arizona.

Chapra, S., & Canale, R. (2014). *Numerical methods for engineers* (7th ed.). New York, USA: McGraw-Hill.

Chen, S. (1995). Nonlinear time series modeling and prediction using Gaussian RBF networks with enhanced clustering and RLS learning. *Electronics Letters, 31*(2), 117–118.

Chen, S., Cowan, C. F. N., & Grant, P. M. (1991). Orthogonal least squares learning algorithm for radial basis function networks. *IEEE Transactions on Neural Networks, 2*(2), 302–309.

Cho, J., Principe, J. C., Erdogmus, D., & Motter, M. A. (2006). Modeling and inverse controller design for an unmanned aerial vehicle based on the self-organizing map. *IEEE Transactions on Neural Networks, 17*(2), 445–460.

Cho, K. C, Ma, Y. B., & Lee, J. S. (2006). Design of computational grid-based intelligence ART-1 classification system for nioinformatics applications. In *Proceedings of the 5th International Conference on Grid and Cooperative Computing Workshops* (pp. 367–370). Changsha, China.

Chowdhury, M., Alouani, A., & Hossain, F. (2010). Comparison of ordinary kriging and artificial neural network for spatial mapping of arsenic contamination of groundwater. *Stochastic Environmental Research and Risk Assessment, 24*(1), 1–7.

Cireşan, D. C., Giusti, A., Gambardella, L. M., & Schmidhuber, J. (2013). Mitosis detection in breast cancer histology images with deep neural networks. *Lecture Notes in Computer Science, 8150*, 411–418.

Coakley, J. R., & Brown, C. E. (2000). Artificial neural networks in accounting and finance: Modeling issues. *International Journal of Intelligent Systems in Accounting, Finance & Management, 9*(2), 119–144.

Curry, B., & Morgan, P. H. (2006). Model selection in neural networks: Some difficulties. *European Journal of Operational Research, 170*(2), 567–577.

D'Inverno, M., & Luck, M. (2004). *Understanding agent systems*. Berlin, Germany: Springer.

Dailey, M. N., & Cottrell, G. W. (2002). EMPATH: A neural network that categorizes facial expressions. *Journal of Cognitive Neuroscience, 14*, 1158–1173.

Dasgupta, D. (2006). Advances in artificial immune systems. *IEEE Computational Intelligence Magazine, 1*(4), 40–49.

Dasgupta, D., & Michalewicz, Z. (1997). *Evolutionary algorithms in engineering applications*. Berlin, Germany: Springer.

Deeb, O. (2010). Correlation ranking and stepwise regression procedures in principal components artificial neural networks modeling with application to predict toxic activity and human serum albumin binding affinity. *Chemometrics and Intelligent Laboratory Systems, 104*(2), 181–194.

Dong, Y., Shao, M., & Tai, X. (2008). An adaptive counter propagation network based on soft competition. *Pattern Recognition Letters, 29*(7), 938–949.

Duda, R. O., Hart, P. E., & Stork, D. G. (2004). *Pattern Classification* (2nd ed.). New York, USA: Wiley.

Elman, J. L. (1990). Finding structure in time. *Cognitive Science, 14*, 179–211.

Faggin, F. (1991). Tutorial Notes: VLSI implementation of neural networks. In *Proceedings of the International Joint Conference on Neural Networks*, Seattle, Washington, USA.

Fernandes, R. A. S., Silva, I. N., & Oleskovicz, M. (2013). Load profile identification interface for consumer online monitoring purposes in smart grids. *IEEE Transactions on Industrial Informatics, 9*(3), 507–1517.

Finnof, W., Hergert, F., & Zimmermann, H. G. (1993). Improving model selection by nonconvergent methods. *Neural Networks, 6*(6), 771–783.

Foresee, F. D., & Hagan, M. T. (1997). Gauss-newton approximation to Bayesian regularization. In: *Proceedings of the International Joint Conference on Neural Networks* (pp. 1930–1935). Hong Kong, China.

García, C., & Moreno, J. A. (2004). The Hopfield associative memory network: Improving performance with the kernel "trick". *Lecture Notes in Computer Science (IBERAMIA'2004)* (Vol. 3315, pp. 871–880).

Glover, F., & Laguna, M. (1998). *Tabu Search*. London, UK: Springer.

Goldberg, D. E. (1989). *Genetic algorithms in search*. Addison Wesley, Reading, Massachusetts, USA: Optimization and Machine Learning.

Goldberg, D. E. (2002). *The design of innovation (genetic algorithms and evolutionary computation)*. Dordrecht, The Netherlands: Kluwer Academic Publishers.

Graham, A. (1986). *Kronecker Products and Matrix Calculus With Applications*. Chichester, West Sussex, UK: Ellis Horwood Ltd. (Reprint Edition).

Grossberg, S. (1974). Classical and instrumental learning by neural networks. *Progress in Theoretical Biology* (Vol. 3, pp. 51–141). New York, USA: Academic Press.

Grossberg, S. (1976a). Adaptive pattern classification and universal recoding: Part I (Parallel development and coding of neural feature detectors). *Biological Cybernetics, 23*(3), 121–134.

Grossberg, S. (1976b). Adaptive pattern classification and universal recoding: Part II (Feedback, expectation, olfaction, and illusions). *Biological Cybernetics, 23*(4), 187–202.

Grossberg, S. (1980). How does a brain build a cognitive code? *Psychological Review, 87*, 1–51.

Hagan, M. T., & Menhaj, M. B. (1994). Training feedforward networks with the Marquardt algorithm. *IEEE Transactions on Neural Networks, 5*(6), 989–993.

Hagan, M. T., Demuth, H. B., Beale, M. H., & Jesús, O. (2014). *Neural network design* (2nd ed.). Oklahoma State University, Stillwater, Oklahoma, USA.

Hampshire II, J.B., & Pearlmutter, B. (1991). Equivalence proofs for multi-layer perceptron classifiers and the Bayesian discriminant function. In *Connectionist Models: Proceedings of the 1990 Summer School, Morgan Kaufmann*, San Mateo, California, USA.

Han, H.-G., Chen, Q.-L., & Qiao, J.-F. (2011). An efficient self-organizing RBF neural network for water quality prediction. *Neural Networks, 24*(7), 717–725.

Hartigan, J. A. (1975). *Clustering Algorithms*. New York, USA: Wiley.

Haykin, S. (2009). *Neural Networks and Learning Machines* (3rd ed.). Upper Saddle River, New Jersey, USA: Prentice Hall.

He, Y., Li, X., & Deng, X. (2007). Discrimination of varieties of tea using near infrared spectroscopy by principal component analysis and BP model. *Journal of Food Engineering, 79* (4), 1238–1242.

Hebb, D. O. (1949). *The organization of behavior: A neuropsychological theory*. New York, USA: Wiley.

Hecht-Nielsen, R. (1987a). Counterpropagation networks. *Applied Optics, 26*(23), 4979–4984.

Hecht-Nielsen, R. (1987b). Counterpropagation networks. In *Proceedings of the IEEE First International Conference on Neural Networks* (Vol. 2, pp. 19–32). San Diego, California.

Hinton, G., Deng, L., Yu, D., Dahl, G. E., Mohamed, A., Jaitly, N., et al. (2012). Deep neural networks for acoustic modeling in speech recognition: The shared views of four research groups. *IEEE Signal Processing Magazine, 29*(6), 82–97.

Hodgkin, A. L., & Huxley, A. F. (1952). A quantitative description of membrane current and its application to conduction and excitation in nerve. *Journal of Physiology, 117*, 500–544.

Hopfield, J. J. (1982). Neural network and physical systems with emergent collective computational abilities. *Proceedings of the National Academy of Sciences of the USA, 79*(8), 2554–2558.

Hopfield, J. J. (1984). Neurons with graded response have collective computational properties like those of two-state neurons. *Proceedings of the National Academy of Sciences of the USA, 81* (10), 3088–3092.

Hopfield, J. J., & Tank, D. W. (1985). Neural computation of decisions in optimization problems. *Biological Cybernetics, 52*(3), 141–152.

Huang, Y., & Yang, X.-S. (2006). Chaos and bifurcation in a new class of simple Hopfield neural network. *Lecture Notes in Computer Science (ISNN'2006), 3971*, 316–321.

Jordan, M. I. (1986). Attractor dynamics and parallelism in a connectionist sequential machine. In *Proceedings of the 8th Annual Conference on Cognitive Science* (pp. 531–546).

Kandel, E. R., Schwartz, J. H., & Jessell, T. M. (2012). *Principles of neural science* (5th ed.). New York, USA: McGraw-Hill.

Karaa, Y., Boyacioglub, M. A., & Baykan, Ö. K. (2011). Predicting direction of stock price index movement using artificial neural networks and support vector machines: The sample of the Istanbul stock exchange. *Expert Systems with Applications, 38*(5), 5311–5319.

Kennedy, J., & Eberhart, R. C. (2001). *Swarm intelligence.* San Diego, California, USA: Morgan Kaufmann Publishers.

Khan, J., Wei, J. S., Ringnér, M., Saal, L. H., Ladanyi, M., Westermann, F., et al. (2001). Classification and diagnostic prediction of cancers using gene expression profiling and artificial neural networks. *Nature Medicine, 7*(6), 673–679.

Kirkpatrick, S., Gelatt, C. D., & Vecchi, M. P. (1983). Optimization by simulated annealing. *Science, 220*(4598), 671–680.

Kohavi, R. (1995). A study of cross-validation and bootstrap for accuracy estimation and model selection. In *Proceedings of the 14th International Joint Conference on Artificial Intelligence* (pp. 1137–1145). Montreal, Canada.

Kohonen, T. (1982). Self-organized formation of topologically correct feature maps. *Biological Cybernetics, 43*, 59–69.

Kohonen, T. (1984). *Self-organization and associate memory.* London, UK: Springer.

Kohonen, T. (1990). Improved versions of learning vector quantization. In *Proceedings of the International Joint Conference on Neural Networks* (pp. 545–550). San Diego, California.

Kohonen, T. (2000). *Self-organizing maps* (3rd ed.). Berlin, Germany: Springer.

Kolmogorov, A. N. (1957). On the representation of continuous functions of several variables by superposition of continuous functions of one variable and addition. *Doklady Akademii. Nauk USSR, 114*, 679–681.

Kurose, J. F., & Ross, K. W. (2016). *Computer Networking—A Top-Down Approach. Pearson* (7th ed.). UK: London.

Lang, K. J., & Hinton, G. E. (1988). *The Development of the Time-Delay Neural Network Architecture for Speech Recognition.* Technical Report CMU-CS-88-152, Carnegie-Mellon University, Pittsburgh, Pennsylvania, USA.

LeCun, Y. (1989). *Generalization and Network Design Strategies.* Technical Report CRG-TR-89-4, Department of Computer Science, University of Toronto, Canada.

Lee, D.-L. (2006). Improvements of complex-valued Hopfield associative memory by using generalized projection rules. *IEEE Transactions on Neural Networks, 17*(5), 1341–1347.

Lee, S.-W., & Kim, J.-S. (1995). Multi-lingual, multi-font and multi-size large-set character recognition using self-organizing neural network. In *Proceedings of the Third International Conference on Document Analysis and Recognition* (pp. 28–33). Montreal, Canada.

Lek, S., & Guégan, J.-F. (2012). *Artificial neuronal networks—Application to ecology and evolution*. Berlin, Germany: Springer.

Leondes, T. L. (2006). *Control and dynamic systems—Neural Network systems techniques and applications*. San Diego, California, USA: Academic Press.

Li, J.-H., Michel, A. N., & Porod, W. (1989). Analysis and synthesis of a class of neural networks: Linear systems operating on a closed hypercube. *IEEE Transactions on Circuits and Systems, 36*(11), 1405–1422.

Lin, Z., Khorasani, K., & Patel, R. V. (1990). A counter-propagation neural network for function approximation. In *Proceedings of the IEEE International Conference on Systems, Man and Cybernetics* (pp. 382–384). Los Angeles, California, USA.

Lippmann, R. P. (1987). An Introduction to computing with neural nets. *IEEE ASSP Magazine, 4*, 4–22.

Ljung, L. (1999). *System Identification: Theory for the user* (2nd ed.). Upper Saddle River, New Jersey, USA: Prentice Hall.

Luenberger, D. G., & Ye, Y. (2016). *Linear and nonlinear programming* (4th ed.). New York, NY, USA: Springer.

Lui, H. C. (1990). Analysis of decision contour of neural network with sigmoidal nonlinearity. In *Proceedings of the International Joint Conference on Neural Networks* (pp. 655–658). Washington/DC, USA.

Makhoul, J., El-Jaroudi, A., & Schwartz, R. (1989). Formation of disconnected decision regions with a single hidden layer. In *Proceedings of the International Joint Conference on Neural Networks* (pp. 455–460). Cambridge, Massachusetts, USA.

Malek, A., & Yari, A. (2005). Primal-dual solution for the linear programming problems using neural networks. *Applied Mathematics and Computation, 167*, 198–211.

McCulloch, W. S., & Pitts, W. (1943). A logical calculus of the ideas immanent in nervous activity. *Bulletin of Mathematical Biophysics, 5*, 115–133.

Mendyk, A., & Jachowicz, R. (2007). Unified methodology of neural analysis in decision support systems built for pharmaceutical technology. *Expert Systems with Applications, 32*(4), 1124–1131.

Michalewicz, Z. (1999). *Genetic algorithms + data structures = evolution programs* (3rd ed.). Berlin, Germany: Springer.

Minsky, M. L., & Papert, S. A. (1969). *Perceptrons: An introduction to computational geometry*. Cambridge, Massachusetts, USA: The MIT Press.

Mishra, D., Chandra Bose, N. S., Tolambiya, A., Dwivedi, A., Kandula, P., Kumar, A., & Kalra, P. K. (2006). Color image compression with modified forward-only counterpropagation neural network: Improvement of the quality using different distance measures. In *Proceedings of the 9th International Conference on Information Technology* (pp. 139–140). Bhubaneswar, India.

Moon, T. K. (2005). *Error correction coding—Mathematical methods and algorithms*. New Jersey, USA: Wiley.

Morns, I. P., & Dlay, S. S. (2003). The DSFPN: A new neural network and circuit simulation for optical character recognition. *IEEE Transactions on Signal Processing, 51*(12), 3198–3209.

Muezzinoglu, M. K., Guzelis, C., & Zurada, J. M. (2005). An energy function-based design method for discrete Hopfield associative memory with attractive fixed points. *IEEE Transactions on Neural Networks, 16*(1), 370–378.

Narendra, K. S., & Parthasarathy, K. (1990). Identification and control of dynamical systems using neural networks. *IEEE Transactions on Neural Networks, 1*(1), 4–27.

Nazario, S. L. S. (2007). *Milk Characterization Using Ultrasonic Techniques and Neural Networks*. Master's Thesis, São Paulo State University at Ilha Solteira, São Paulo, Brazil.

Nelles, O. (2010). *Nonlinear system identification: From classical approaches to neural networks and fuzzy models*. Berlin, Germany: Springer.

Norgaard, M., Ravn, O., Poulsen, N. K., & Hansen, L. K. (2010). *Neural Networks for Modelling and Control of Dynamic Systems* (2nd ed.). Germany: Springer.

Nowlan, S. J. (1989). Maximum likelihood competitive learning. *Advances in neural information processing systems 2* (pp. 574–582). San Mateo, California, USA: Morgan Kaufmann.

Ortega, A. V., Silva, I. N. (2008). Neural network model for designing automotive devices using SMD LED. *International Journal of Automotive Technology), 9*(2), 203–210.

Otsu, N. (1979). A threshold selection method from grey-level histograms. *IEEE Transactions on Systems, Man and Cybernetics, 9*(1), 62–66.

Park, J., & Sandberg, I. W. (1991). Universal approximation using radial-basis-function networks. *Neural Computation, 3*, 246–257.

Parker, J. R. (2010). *Algorithms for image processing and computer vision* (2nd ed.). New York, USA: Wiley.

Parsons, S., & Jones, G. (2000). Acoustic identification of twelve species of echolocating bat by discriminant function analysis and artificial neural networks. *Journal of Experimental Biology, 203*(17), 2641–2656.

Pedrycz, W., & Gomide, F. (2007). *Fuzzy systems engineering—Toward human-centric computing.* Hoboken, New Jersey, USA: Wiley.

Pincus, M. (1970). A Monte Carlo method for the approximate solution of certain types of constrained optimization problems. *Operation Research, 18*, 1225–1228.

Powell, M. J. D. (1987). Radial basis functions for multivariable interpolation: A review. *Algorithms for Approximation* (pp. 143–167). Oxford, UK: Oxford University Press.

Rajasekaran, S., & Raj, R. A. (2004). Image recognition using analog-ART1 architecture augmented with moment-based feature extractors. *Neurocomputing, 56*, 61–77.

Reed, R. D., & Marks, R. J. I. I. (1999). *Neural smithing: Supervised learning in feedforward artificial neural networks.* Cambridge, Massachusetts, USA: MIT Press.

Richter, T., Oliveira, A. F., & Silva, I. N. (2010). Virtual oxygen sensor implementation using artificial neural networks. In *Technological Developments in Education and Automation* (Chap. 41). New York, USA: Springer.

Riedmiller, M., & Braun, H. (1993). A direct adaptive method for faster backpropagation learning: the RPROP algorithm. In *Proceedings of the International Joint Conference on Neural Networks* (pp. 586–591). San Francisco, California, USA.

Ripley, B. D. (1996). *Pattern recognition and neural networks.* Cambridge, UK: Cambridge University Press.

Ritter, H., Martinetz, T., & Schulten, K. (1992). *Neural computation and self-organizing maps— An introduction.* New York, USA: Addison-Wesley.

Rosenblatt, F. (1958). The perceptron: A probabilistic model for information storage and organization in the brain. *Psychological Review, 65*, 386–408.

Ross, T. J. (2010). *Fuzzy logic with engineering applications* (3rd ed.). Chichester, West Sussex, UK: Wiley.

Rumelhart, D. E., Hinton, G. E., & Williams, R. J. (1986). Learning internal representations by error propagation. In *Parallel Distributed Processing* (Vol. 1, Chap. 8). Cambridge, Massachusetts, USA: MIT Press.

Shepherd, G. M. (2004). *The synaptic organization of the brain* (5th ed.). USA: Oxford University Press, New York.

Silva, C. B. S. (2007). *Nuclear magnetic resonance signal processing using neural classifier for recognition of beef (in portuguese).* Master's Thesis, University of São Paulo at São Carlos, São Paulo, Brazil.

Silva, I. N. (1997). *A Neuro-fuzzy Approach to Systems Optimization and Robust Identification.* Ph.D. Thesis, University of Campinas, São Paulo, Brazil.

Silva, I. N., Amaral, W. C., & Arruda, L. V. R. (2007). A novel approach based on recurrent neural networks applied to nonlinear systems optimization. *Applied Mathematical Modelling, 31*, 78–92.

Silva, I. N., Arruda, L. V. R., & Amaral, W. C. (2001). An efficient model of neural networks for dynamic programming. *International Journal of Systems Science, 32*(6), 715–722.

Sontag, E. D. (1992). Feedback stabilization using two-hidden-layer nets. *IEEE Transactions on Neural Networks, 3*(6), 981–990.

Srinivasan, A., & Batur, C. (1994). Hopfield/ART-1 neural network-based fault detection and isolation. *IEEE Transactions on Neural Networks, 5*(6), 890–899.

Steinera, N. Y., Hisselb, D., Moçotéguy, Ph, & Candussoc, D. (2011). Diagnosis of polymer electrolyte fuel cells failure modes (flooding & drying out) by neural networks modeling. *International Journal of Hydrogen Energy, 36*(4), 3067–3075.

Steinhaus, H. D. (1956). Sur la division des corp materiels en parties. *Bulletin L'Academie Polonaise des Science, IV*(C1-III), 801–804; Sutton, R. S., & Barto, A. G. (1998). *Reinforcement learning: An introduction.* Cambridge, Massachusetts, USA: MIT Press.

Suykens, K., Vandewalle, J. P. L., & De Moor, B. L. (2001). *Artificial neural networks for modelling and control of non-linear systems.* Berlin, Germany: Springer.

Sygnowski, W., & Macukow, B. (1996). Counter-propagation neural network for image compression. *Optical Engineering, 35*(8), 2214–2217.

Tabach, E. E., Lancelot, L., Shahrour, I., & Najjar, Y. (2007). Use of artificial neural network simulation metamodelling to assess groundwater contamination in a road project. *Mathematical and Computer Modelling, 45*(7–8), 766–776.

Tank, D. W., & Hopfield, J. J. (1986). Simple 'neural' optimization networks: an A/D converter, signal decision circuit, and a linear programming circuit. *IEEE Transactions on Circuits and Systems, 33*(5), 533–541.

Tateyama, T., & Kawata, S. (2004). Plant layout planning using ART neural networks. In *Proceedings of the SICE 2004 Annual Conference* (pp. 1863–1868). Sapporo, Japan.

Tian, H., & Shang, Z. (2006). Artificial neural network as a classification method of mice by their calls. *Ultrasonics, 44,* e275–e278.

Too, G.-P. J., Chen, S. R., & Hwang, S. (2007). Inversion for acoustic impedance of a wall by using artificial neural network. *Applied Acoustics, 68*(4), 377–389.

Tsoukalas, L. H., & Uhrig, R. E. (1997). *Fuzzy and neural approaches in engineering.* New York, USA: Wiley.

Vakil-Baghmisheh, M.-T., & Pavesic, N. (2003). Premature clustering phenomenon and new training algorithms for LVQ. *Pattern Recognition, 36*(8), 1901–1912.

Vapnik, V. N. (1998). *Statistical learning theory.* New York, USA: Wiley.

Vicente, B. G. L. Z., Cezare, M. J., & Silva, I. N. (2007). Neural controller to idle in internal combustion engines (in portuguese). In *Proceedings of the Intelligent Automation Brazilian Symposium* (pp. 1–6). Florianópolis, Santa Catarina, Brazil.

Vidyasagar, M. (2002). *Nonlinear systems analysis* (2nd ed.). Philadelphia, Pennsylvania, USA: SIAM Publications.

Waibel, A., Hanazawa, T., Hinton, G., Shikano, K., & Lang, K. (1989). Phoneme recognition using time-delay neural networks. *IEEE Transactions on Acoustics, Speech and Signal Processing, 37*(3), 328–339.

Wang, J. (2004). A recurrent neural network for solving the shortest path problem. *IEEE Transactions on Circuits and Systems—Part I, 43*(6), 482–486.

Wasserman, P. D. (1989). *Neural computing—Theory and practice.* New York, USA: Van Nostrand Reinhold.

Watkins, C. J. C. H. (1989). *Learning from Delayed Rewards.* Ph.D. Thesis, Psychology Department, Cambridge University, UK.

Werbos, P.J. (1974). *Beyond Regression: New Tools for Prediction and Analysis in the Behavioral Sciences.* Ph.D. Thesis, Havard University, Cambridge, Massachusetts, USA.

Widrow, B., & Hoff, M. E. (1960). Adaptive switching circuits. In *Proceedings of the IRE WESCON Convention Record* (pp. 96–104).

Widrow, B., & Winter, R. (1988). Neural nets for adaptive filtering and adaptive pattern recognition. *IEEE Computer Magazine, 21*(3), 25–39.

Wirth, N. (1989). *Algorithms and data structures.* Upper Saddle River, New Jersey, USA: Prentice-Hall.

Witten, I. H., Flank, E., & Hall, M. A. (2011). *Data mining—Practical machine learning tools and techniques. morgan kaufmann* (3rd ed.). California, USA: San Francisco.

Xia, Y., & Wang, J. (2004). A recurrent neural network for nonlinear convex optimization subject to nonlinear inequality constraints. *IEEE Transactions on Circuits and Systems—Part I, 51*(7), 1385–1394.

Xiang, C., Ding, S. Q., & Lee, T. H. (2005). Geometrical interpretation and architecture selection of MLP. *IEEE Transactions on Neural Networks, 16*(1), 84–96.

Yan, H., Jiang, Y., Zheng, J., Peng, C., & Li, Q. (2006). A multilayer perceptron-based medical decision support system for heart disease diagnosis. *Expert Systems with Applications, 30*(2), 272–281.

Yang, C.-H., Luo, C.-H., Yang, C.-H., & Chuang, L.-Y. (2004). Counter-propagation network with variable degree variable step size LMS for single switch typing recognition. *Bio-medical Materials and Engineering, 14*(1), 23–32.

Young, R. J., Ritthaler, M., Zimmer, P., McGraw, J., Healy, M. J., & Caudell, T. P. (2007). Comparison of adaptive resonance theory neural networks for astronomical region of interest detection and noise characterization. In *Proceedings of the International Joint Conference on Neural Networks* (pp. 2123–2128), Albuquerque, New Mexico, USA.

Zadeh, L. (1992). Fuzzy logic, neural networks and soft computing. In *Proceedings of the 2nd International Conference on Fuzzy Logic and Neural Networks* (pp. 13–14). Iizuka, Japan.

Zhang, Q., & Stanley, S. J. (1999). Real-time water treatment process control with artificial neural networks. *Journal of Environmental Engineering, 125*(2), 153–160.

Zhang, Q., & Xu, X. W. J. (2007). Delay-dependent global stability results for delayed Hopfield neural networks. *Chaos, Solitons & Fractals, 34*(2), 662–668.

Zhang, Q.-B., Hebda, R. J., Zhang, Q.-J., & Alfaro, R. I. (2000). Modeling tree-ring growth responses to climatic variables using artificial neural networks. *Journal of Forest Science, 46* (2), 229–239.

Zhang, Z., & Friedrich, K. (2003). Artificial neural networks applied to polymer composites: a review. *Elsevier Science, Composites Science and Technology, 63*(14), 2029–2044.

Index

A
Activation functions
bipolar step function, 13
Gaussian function, 17
hyperbolic tangent function, 16
linear function, 18
logistic function, 15
step function, 13
symmetric ramp function, 14
ADALINE, 6, 7, 22, 41–54, 56, 62, 69
Adaptive resonance principle, 192, 205
Adaptive resonance theory, 7, 189, 230
ANN applications
associative memory, 8
data clustering, 8
pattern classification, 8
prediction systems, 8
process control, 8
system optimization, 8
universal curve fitting, 8
ANN architectures
mesh architecture, 24
multiple-layer feedforward architecture, 23
recurrent feedforward architecture, 24
single-layer feedforward architecture, 22
ANN layers
hidden, intermediate, or invisible layers, 22
input layer, 21
output layer, 22
ART-1
comparison phase, 196
parameter initialization phase, 193
provision phase, 201
recognition phase, 195
search phase, 198
update, 199
Artificial immunologic systems, 5
Artificial neurons
activation function, 12

activation potential, 12
activation threshold, 12
input signals, 12
linear aggregator, 12
output signal, 12
synaptic weights, 12
Artificial synapses, 5
Associative memories, 140, 145, 146, 148,
150, 151, 153, 154

B
Backpropagation, 7, 55, 57–59, 61–66, 68, 71,
73, 75, 78, 79, 88, 109, 110, 112, 114,
157
Bayesian boarders, 181
Biological neuron, 9, 18

C
Chaotic behavior, 142
Clustering method, 121
Cognitive-emotional interactions, 189
Combinatorial optimization, 140, 154
Comparison layer, 191, 193
Competition mechanism, 169
Competitive learning process, 158
Computational intelligence, 5
Connectionist systems, 5
Contextual map, 168, 169
Convergence process, 46, 47, 49, 51, 52, 69,
71–74, 109, 140, 153, 162, 170, 181,
183, 184, 270
Convex region, 81–83, 109
Counter-propagation architecture, 182
Cross-validation methods, 97, 100, 103, 104

D
Data clustering, 24, 185, 190
Data organization, 6

© Springer International Publishing Switzerland 2017
I.N. da Silva et al., *Artificial Neural Networks*,
DOI 10.1007/978-3-319-43162-8

Printed in the United States
By Bookmasters